NATURE-INSPIRED METHODS FOR STOCHASTIC, ROBUST AND DYNAMIC OPTIMIZATION

Edited by **Javier Del Ser** and **Eneko Osaba**

Nature-inspired Methods for Stochastic, Robust and Dynamic Optimization
http://dx.doi.org/10.5772/intechopen.71401
Edited by Javier Del Ser and Eneko Osaba

Contributors

Jose Garcia, Alvaro Peña, Maxim Dulebenets, Siew Mooi Lim, Kuan Yew Leong, Javier Del Ser Lorente, Eneko Osaba

Notice

Statements and opinions expressed in the chapters are these of the individual contributors and not necessarily those of the editors or publisher. No responsibility is accepted for the accuracy of information contained in the published chapters. The publisher assumes no responsibility for any damage or injury to persons or property arising out of the use of any materials, instructions, methods or ideas contained in the book.

First published in London, United Kingdom, 2018 by IntechOpen
IntechOpen is the global imprint of INTECHOPEN LIMITED, registered in England and Wales, registration number: 11086078, The Shard, 25th floor, 32 London Bridge Street
London, SE19SG – United Kingdom
Printed in Croatia

British Library Cataloguing-in-Publication Data
A catalogue record for this book is available from the British Library

Additional hard copies can be obtained from orders@intechopen.com

Nature-inspired Methods for Stochastic, Robust and Dynamic Optimization, Edited by Javier Del Ser and Eneko Osaba
p. cm.
Print ISBN 978-1-78923-328-5
Online ISBN 978-1-78923-329-2

We are IntechOpen,
the world's leading publisher of
Open Access books
Built by scientists, for scientists

3,600+
Open access books available

113,000+
International authors and editors

115M+
Downloads

Our authors are among the

151
Countries delivered to

Top 1%
most cited scientists

12.2%
Contributors from top 500 universities

CLARIVATE ANALYTICS
BOOK
CITATION
INDEX
INDEXED

WEB OF SCIENCE™

Selection of our books indexed in the Book Citation Index
in Web of Science™ Core Collection (BKCI)

Interested in publishing with us?
Contact book.department@intechopen.com

Numbers displayed above are based on latest data collected.
For more information visit www.intechopen.com

Meet the editors

Prof. Dr. Javier Del Ser received his first PhD degree in Telecommunication Engineering from the University of Navarra, Spain, and his second PhD degree in Computational Intelligence from the University of Alcala, Spain. He is a principal researcher in data analytics and optimization at Tecnalia, a visiting fellow at the Basque Center for Applied Mathematics (BCAM) and a part-time lecturer at the University of the Basque Country (UPV/EHU). His research interests gravitate on the use of descriptive, prescriptive and predictive algorithms for data mining and optimization in a diverse range of application fields. He has published more than 190 articles, co-supervised 6 PhD degree theses, edited 4 books, coauthored 6 patents and participated/led more than 35 research projects.

Dr. Eneko Osaba received his B.Sc. and M.Sc. degrees in Computer Science from the University of Deusto. He obtained his PhD degree in Artificial Intelligence in 2015 from the same university, being the recipient of a Basque Government doctoral grant. He has participated in the proposal, development and justification of 15 research projects. He has participated in the development of more than 70 papers, having JCR impact factor for 20 of them. He has performed several stays in universities of the United Kingdom, Italy and Malta. He served as a member of the program and/or organizing committee in more than 15 international conferences, and he is a member of the editorial board of the International Journal of Artificial Intelligence.

Contents

Introductory Chapter: Nature-Inspired Methods for Stochastic, Robust, and Dynamic Optimization

Eneko Osaba and Javier Del Ser

Additional information is available at the end of the chapter

http://dx.doi.org/10.5772/intechopen.78009

1. Introduction

Optimization is one of the most studied fields in the wide field of artificial intelligence. Hundreds of studies published year after year focus on solving many diverse problems of this kind by resorting to a vast spectrum of solvers. Within this class of problems, several problem flavors can be identified depending on the characteristics of their constituent fitness functions and support of their optimization variables, such as linear, continuous or combinatorial. Efficiently tackling such optimization problems requires huge computational resources, especially when the formulated problem at hand represents complex real-world situations with hundreds of variables and constraints. For these reasons and due to the inherently practical utility of optimization algorithms, very heterogeneous problem-solving approaches have been developed by the community over the last decades for their application to these problems. From a general perspective, optimization methods can be classified as exact, heuristics, and metaheuristics. In this chapter, the focus is placed on the latter two families, in particular in those algorithmic variants where biological processes observed in nature have lied at the motivating core of the operators underlying their search mechanisms. In other words, we will center our attention on Nature-Inspired methods for efficient optimization and problem solving.

In this context, Nature-Inspired algorithms have recently gained ever-growing popularity in the community, with an unprecedented body of the literature related to assorted algorithmic approaches suited to deal with problem formulations by leveraging the self-learning capability of their mimicked natural phenomena. The rationale behind the momentum acquired by this broad family of methods lies in their outstanding performance, which has hitherto been evinced in hundreds of research fields and problem scenarios. In this regard, many different inspirational sources have been proposed for constructing optimization methods, such as the behavioral patterns of bats [1], fireflies [1], bees [2] or the stigmergy by which ants communicate to each other when looking over an area for a food source [3], which add to the mechanisms behind genetic inheritance that stimulated the advent of the seminal branch of genetic algorithms [4].

In recent years, most of these Nature-Inspired methods have been successfully applied to a wide variety of topics. To cite a few, the aforementioned Bat algorithm has been applied to problems related to energy [5], sports training planning [6] or logistics [7, 8], whereas the Firefly Algorithm has been applied to selected applications in medicine [9], job-shop scheduling [10] or goods distribution, and logistics [11, 12]. This is a very reduced yet exemplary bibliographic sample of the heterogeneous research activity around Nature-Inspired methods. A thorough review of the state of the art in this topic can be extensive, reason for which many comprehensive surveys have been lately contributed to reflect the huge literature produced around certain algorithms. Nevertheless, Genetic Algorithms and Ant Colony Systems are, arguably, the most widely resorted algorithms of this kind, with recent literature compendiums focused on these both approaches appearing in the literature on a yearly basis [13, 14].

This introductory chapter contributes to this line of research by presenting applications of Nature-Inspired solvers to three specific branches of optimization problems, namely, stochastic, dynamic, and robust optimization. We next provide a more elaborated presentation of each of such branches.

2. Dynamic optimization

In optimization problems, it is often the case that the parameters based on which fitness function(s) and constraints are defined remain unaltered over the period of time in which the solution obtained by the solver is considered to be optimal. Therefore, such parameters are assumed to be known a priori and fixed from the very beginning of the problem solving process. In dynamic optimization, however, this stability condition may not hold, this one or more constraints and/or fitness function of the problem can vary dynamically along time, even after the problem is solved and the solution is applied. The setup can be even more involved if new parameters appear at any step of the process, which must not only be included in the problem formulation but also accommodated by the technique at hand. Due to these exceptional situations, this casuistry demands efficient algorithmic means to solve optimization problems in an on-line fashion.

Dynamism in any aspect of the problem is a practical circumstance that emerges in almost any field where the context of the problem evolves along time due to exogenous factors to the initially formulated problem statement. One of the scenarios, where dynamic optimization is under active investigation, is transportation and mobility, in which the dynamism of the considered parameters can force re-planning previously traced routes, even if the vehicle is already on the road. This hypothesized case can be produced either by the appearance of any incident over the road network or the arrival of unexpected information that was not present when the initial route optimization was performed. An example of this kind of problem was presented in [15], in which a vehicle routing problem is modeled by integrating the information about future customers' dynamic requests. Another problem prone to considering this characteristic is the job-shop scheduling problem and its multiple variants, as can be seen in recently published studies such as [16, 17]. Several interesting surveys are available on this topic, such as [18], in which the application of swarm intelligence methods to dynamic optimization is reviewed. In [19], on the other hand, Evolutionary Algorithms are analyzed for the same class of optimization problems.

3. Stochastic optimization

Stochastic optimization is another problem variant that finds its motivation in real application scenarios. This class of optimization problems can be defined as the process of maximizing or minimizing the value of a mathematical or statistical function, in which one or more of its values are subject to randomness. This stochastic nature may involve random objective functions and/or random restrictions, which ease the modeling of real-world problems subject to non-negligible sources of uncertainty, imprecision or randomness.

The need for stochastic optimization techniques emerge from a wide variety of real-world problems related to business analytics, electrical power production or energy management, among many others. In [20], for example, the so-called unit commitment problem is endowed with this feature to model and handle the uncertainty of the electric power generation process in the scheduling and dispatching of the produced energy. On the other hand, the authors in [21] regard power system management as a stochastic optimization problem, considering microgrids capable of controlling their local generation and demand with the presence of an uncertain amount of generated renewable energy.

Focused on Nature-Inspired techniques, examples such as the one found in [22] are worth to be mentioned. In this work, a Firefly Algorithm is used to tackle a multi-objective active/reactive power dispatch problem, with the existence of wind generation and load uncertainties. Another example can be accessed in [23], in which a Genetic Algorithm is utilized for efficiently solving a condition-based maintenance optimization problem subject to uncertainties.

4. Robust optimization

The third class of optimization problems targeted by this chapter is robust optimization, which denotes a branch of problems where one or more variables that compose the problem is also subject to uncertainty. In this case, however, the scope is placed on the robustness of the produced solutions against the variability of the constraints affected by uncertainty (e.g., the target is always placed on fulfilling simultaneously all constraints disregarding the statistical variability of the problem), as opposed to stochastic optimization which aim at satisfying the constraints up to a prescribed level of probability. This being said, different types of robust optimization problems can be modeled depending on how extreme values for the variable parameters are formulated. One of these types is referred to as local robustness [24], where a measure of robustness is designed to accommodate small perturbations with respect to the nominal value of the parameter that undergoes stochastic variability. On the other hand, probabilistically robust optimization models [25] quantify the uncertainty in the real value of the parameter of interest using a probability distribution function. Additional classifications are global robustness [24], or non-probabilistic robust optimization models [26].

As has been pointed along this introduction, uncertainty is present in lots of real-world situations. For this reason, robust optimization has also been frequently used for modeling a wide variety of real problems, belonging to different knowledge areas, such as supply chain network design [27] or food distribution [28].

5. Conclusions

This introductory chapter highlights the potential that Nature-Inspired solvers may bring to stochastic, robust, and dynamic optimization problems. Nature has learned from itself from the very beginning of Earth, with manifold processes and intelligent behaviors that have naturally evolved over ages to attain high levels of adaptability and efficiency. It is now time for researchers, lecturers, and practitioners interested in Nature-Inspired optimization to shift their target and span the application of this algorithmic branch to these optimization problems, far less studied so far by the community than other formulated optimization problems.

Author details

Eneko Osaba[1]* and Javier Del Ser[1,2,3]

*Address all correspondence to: eneko.osaba@tecnalia.com

1 TECNALIA Research and Innovation, Derio, Bizkaia, Spain

2 University of the Basque Country (UPV/EHU), Bilbao, Bizkaia, Spain

3 Basque Center for Applied Mathematics (BCAM), Bilbao, Bizkaia, Spain

References

[1] Yang XS. A new metaheuristic bat-inspired algorithm. In: Nature Inspired Cooperative Strategies for Optimization (NICSO 2010). Springer; 2010. pp. 65-74

[2] Karaboga D, Basturk B. A powerful and efficient algorithm for numerical function optimization: Artificial bee colony (abc) algorithm. Journal of Global Optimization. 2007;39(3): 459-471

[3] Dorigo M, Birattari M. Ant colony optimization. In: Encyclopedia of Machine Learning. Springer; 2011. pp. 36-39

[4] Holland JH. Genetic algorithms. Scientific American. 1992;**267**(1):66-73

[5] Kaced K, Larbes C, Ramzan N, Bounabi M, Elabadine Dahmane Z. Bat algorithm based maximum power point tracking for photovoltaic system under partial shading conditions. Solar Energy. 2017;**158**:490-503

[6] Fister I, Rauter S, Yang XS, Ljubič K. Planning the sports training sessions with the bat algorithm. Neurocomputing. 2015;**149**:993-1002

[7] Osaba E, Yang XS, Fister I, Del Ser J, Lopez-Garcia P, Vazquez-Pardavila AJ. A discrete and improved bat algorithm for solving a medical goods distribution problem with pharmacological waste collection. Swarm and Evolutionary Computation. 2018. https://doi.org/10.1016/j.swevo.2018.04.001

[8] Osaba E, Carballedo R, Yang XS, Fister I Jr, Lopez-Garcia P, Del Ser J. On efficiently solving the vehicle routing problem with time windows using the bat algorithm with random reinsertion operators. In: Nature-Inspired Algorithms and Applied Optimization. Springer; 2018. pp. 69-89

[9] Dey N, Samanta S, Chakraborty S, Das A, Chaudhuri SS, Suri JS. Firefly algorithm for optimization of scaling factors during embedding of manifold medical information: An application in ophthalmology imaging. Journal of Medical Imaging and Health Informatics. 2014;**4**(3):384-394

[10] Karthikeyan S, Asokan P, Nickolas S, Page T. A hybrid discrete firefly algorithm for solving multi-objective flexible job shop scheduling problems. International Journal of Bio-Inspired Computation. 2015;**7**(6):386-401

[11] Osaba E, Yang XS, Diaz F, Onieva E, Masegosa AD, Perallos A. A discrete firefly algorithm to solve a rich vehicle routing problem modelling a newspaper distribution system with recycling policy. Soft Computing. 2017;**21**(18):5295-5308

[12] Del Ser J, Torre-Bastida AI, Lana I, Bilbao MN, Perfecto C. Nature-inspired heuristics for the multiple-vehicle selective pickup and delivery problem under maximum profit and incentive fairness criteria. In: IEEE Congress on Evolutionary Computation (CEC), 2017. IEEE; 2017. pp. 480-487

[13] Karakatič S, Podgorelec V. A survey of genetic algorithms for solving multi depot vehicle routing problem. Applied Soft Computing. 2015;**27**:519-532

[14] Afshar A, Massoumi F, Afshar A, Mariño MA. State of the art review of ant colony optimization applications in water resource management. Water Resources Management. 2015;**29**(11):3891-3904

[15] Barkaoui M. A co-evolutionary approach using information about future requests for dynamic vehicle routing problem with soft time windows. Memetic Computing. 2017:1-13. https://doi.org/10.1007/s12293-017-0231-8

[16] Shahrabi J, Adibi MA, Mahootchi M. A reinforcement learning approach to parameter estimation in dynamic job shop scheduling. Computers & Industrial Engineering. 2017; **110**:75-82

[17] Hosseinabadi AAR, Siar H, Shamshirband S, Shojafar M, Nasir MHNM. Using the gravitational emulation local search algorithm to solve the multi-objective flexible dynamic job shop scheduling problem in small and medium enterprises. Annals of Operations Research. 2015;**229**(1):451-474

[18] Mavrovouniotis M, Li C, Yang S. A survey of swarm intelligence for dynamic optimization: Algorithms and applications. Swarm and Evolutionary Computation. 2017;**33**:1-17

[19] Yang S. Evolutionary computation for dynamic optimization problems. In: Conference on Genetic and Evolutionary Computation Proceedings. ACM; 2015. pp. 629-649

[20] Zheng QP, Wang J, Liu AL. Stochastic optimization for unit commitmenta review. IEEE Transactions on Power Systems. 2015;**30**(4):1913-1924

[21] Wang S, Gangammanavar H, Ekşioğlu SD, Mason SJ. Stochastic optimization for energy management in power systems with multiple microgrids. IEEE Transactions on Smart Grid. 2017. https://doi.org/10.1109/TSG.2017.2759159

[22] Liang RH, Wang JC, Chen YT, Tseng WT. An enhanced firefly algorithm to multi-objective optimal active/reactive power dispatch with uncertainties consideration. International Journal of Electrical Power & Energy Systems. 2015;**64**:1088-1097

[23] Compare M, Martini F, Zio E. Genetic algorithms for condition-based maintenance optimization under uncertainty. European Journal of Operational Research. 2015;**244**(2):611-623

[24] Sivaganesan S. Global and local robustness approaches: Uses and limitations. In: Robust Bayesian Analysis. Springer; 2000. pp. 89-108

[25] Beyer HG, Sendhoff B. Robust optimization–A comprehensive survey. Computer Methods in Applied Mechanics and Engineering. 2007;**196**(33–34):3190-3218

[26] Guo SX, Lu ZZ. A non-probabilistic robust reliability method for analysis and design optimization of structures with uncertain-but-bounded parameters. Applied Mathematical Modelling. 2015;**39**(7):1985-2002

[27] Zokaee S, Jabbarzadeh A, Fahimnia B, Sadjadi SJ. Robust supply chain network design: An optimization model with real world application. Annals of Operations Research. 2017; **257**(1–2):15-44

[28] Orgut IS, Ivy JS, Uzsoy R, Hale C. Robust optimization approaches for the equitable and effective distribution of donated food. European Journal of Operational Research. 2018; **269**(2):516-531

Chapter 2

Robust Optimization: Concepts and Applications

José García and Alvaro Peña

Additional information is available at the end of the chapter

http://dx.doi.org/10.5772/intechopen.75381

Abstract

Robust optimization is an emerging area in research that allows addressing different optimization problems and specifically industrial optimization problems where there is a degree of uncertainty in some of the variables involved. There are several ways to apply robust optimization and the choice of form is typical of the problem that is being solved. In this paper, the basic concepts of robust optimization are developed, the different types of robustness are defined in detail, the main areas in which it has been applied are described and finally, the future lines of research that appear in this area are included.

Keywords: optimization, robustness, uncertainly, uncertainty modeling

1. Introduction

Nowadays, using the technologies and techniques associated with the Internet of things, Big Data and artificial intelligence, we are able to capture and process enormous and varied volumes of data. Examples of the above can be observed in different disciplines such as transport [1, 2], mining [3] and agriculture [4] among others. However, in the area of optimization, many problems still work at the level of reference instances [5–8]. To solve real optimization problems, we must consider that these are generally multi-variable problems with restrictions and trade-off between them. In many instances, when a problem is modeled, a point that is not taken into consideration is the uncertainty to which the system is subject. In this sense, our solution can be submitted to questions of the type: How feasible is this solution according to the different scenarios? What is the optimality of this solution? How strict should the treatment of uncertainty be? One way to approach uncertainty is to consider the robustness of the solution. However, the definition of robustness is not trivial and there are several definitions. Ideally, you want to get the best solution and also the most robust one but usually there is a trade-off between these two concepts [9]. Due to the importance and particularity for each problem of this trade-off between the quality and robustness of the solution, a series of definitions have been generated and a series of methods developed to adequately address or estimate the trade-off [10, 11].

Because each problem has its own level of demand regarding the quality of the solution and its treatment with respect to its robustness, it is difficult to provide a single definition of robustness. In some cases, our solution could be considered robust if under certain conditions of the search space or under certain operational conditions, the solution behaves reasonably with respect to its quality, feasibility or optimality. Under other conditions where the management of uncertainty is very strict, the most appropriate result is associated with scenarios that consider the worst case [12].

On the other hand, it has been methodologically argued [13] that instead of transforming and solving the optimization problem with uncertainty in a robust problem, this can be solved in two stages considering the robustness as a separate objective [13]. The argument is based on the fact that a separate analysis allows obtaining more information and understanding about the solution and its robustness, facilitating the decision-making process. On the other hand, considering robustness as part of the problem has advantages over implementation, computational cost and alternatives to solve the problem. In the latter case, modeling the choice of scenarios and the measure of robustness is essential [14].

The aforementioned discussion indicates that the concepts of robustness are still in the process of maturation and there is no clear methodology on how to address robust problems. There are conceptual, computational and application challenges in the area of robust optimization. Usually, the few state of the art reviews found about robust optimization, focus on identifying what areas and types of problems have been addressed. In this article, as a starting point, we present a collection of the different definitions and models with which robust optimization problems have been addressed. Knowledge of the different models used in robust optimization is essential for a proper understanding of the field. Once the main concepts are defined, we proceed to provide an update on the main robust optimization works that have been carried out over the last few years. In Section 2, we will describe the basic concepts associated with uncertainty. We will describe the main robustness models in Section 3. Finally, in Section 4, we will describe the main areas of application.

2. Fundamental concepts

Suppose an engineer who must make constant decisions and face the difficulty of multidimensional problems with some degree of ambiguity or errors in the parameters to analyze and some kind of stochastic uncertainty of the process and its environment. Then this engineer must also determine if the proposed solution is robust. This means that the solution is feasible to apply for any parameter scenario and stochastic uncertainty and that this feasibility remains close to the optimality condition. Then two fundamental concepts appear: the uncertainty of the feasibility of the solution and the uncertainty of the objective value of the function.

2.1. Uncertainty in the feasibility of the solution

Ideally, the engineer would like his or her solution to be feasible for any value of the parameters analyzed; however, this feasibility has consequences. The first consequence corresponds to having a significant computational cost when considering all the possible parameter values. The second consequence is related to the deterioration in the quality of the solution. The more demanding it is with regard to the feasibility of the parameters, the greater the probability of moving away from optimality. Therefore, there is a trade-off, which is related to the problem that is being solved. Then solutions in the area of control theory related to equipment failures should be much stricter regarding the feasibility of the solution than solutions obtained in marketing areas where the effect on a set of clients is not so critical. Therefore, the choice of the uncertainty set plays a fundamental role in the feasibility of solving the problem and in the quality of the solution obtained.

2.2. Uncertainty in the optimality of the solution

It may happen that depending on the set of uncertainty chosen, the optimality of the solution is altered. In this case, robust optimization tries to obtain a solution that performs adequately in the different scenarios; however, all scenarios do not require the same treatment with respect to optimality. Due to the above, in the literature, we can find different concepts of robustness; among the most mentioned are: strict robustness [15], cardinality constrained robustness [16], adjustable robustness [17], lightweight robustness [18], soft robustness [19], lexicographic robustness and regret robustness.

2.3. Uncertainty in the optimization problem

Each real optimization problem suffers from some type of uncertainty that are mainly caused by uncertainty at the level of the measurements or by uncertainties due to changes in the environment of the system. The first case we will refer to microscopic uncertainties and the second will be macroscopic. The optimization problem can be approached in a standard way through a nominal scenario which would describe, for example, the most typical case or an average case. However, in general, the most probable scenario is not trivial to obtain and for some problems, having a more frequent scenario is not the natural way to approach the problem [20]. An optimization problem with constraints can be formally written as shown in Eq. (1).

$$minf(x)$$
$$\text{s.t.} \quad F(x) \le 0 \qquad (1)$$
$$x \in S,$$

where $F : \mathbb{R}^n \to \mathbb{R}^m$ describes a problem of n dimensions with m constraints. $f : \mathbb{R}^n \to \mathbb{R}$ is the objective function and $S \subset \mathbb{R}^n$ is the search space. Our next step is to formalize the uncertainty in the optimization problem. Suppose $\xi \in \mathbb{R}^k$ corresponds to a scenario that could occur in our real problem. Hence, our optimization problem considering the uncertainty scenario ξ, is written in Eq. (2):

$$minf(x, \xi)$$
$$\text{s.t.} \quad F(x, \xi) \leq 0 \qquad (2)$$
$$x \in S,$$

In most problems, it is not known exactly what the value of ξ is, but if it is clear that the problem falls on an uncertainty set $\mathcal{U} \in \mathbb{R}^k$, which represents the scenarios that are enough to consider. Then we have a family of optimization problems given by the pair $(P(\xi), \xi \in \mathcal{U})$. A fundamental objective of robust optimization corresponds to turn this family of problems into a single problem of optimization, where the choice of the set of uncertainty is fundamental for the result and complexity of the problem. For an adequate treatment of the problem of uncertainty, it is fundamental to give structure to the set $\mathfrak{U}$. In literature, it is common to find the following types:

1. Finite uncertainty $\mathfrak{U} = \{\xi^1, \dots \xi^l\}$.

2. Interval-based uncertainty $\mathfrak{U} = |\xi_1, \widehat{\xi}_1| \times \dots \times |\xi_k, \widehat{\xi}_k|$.

3. Norm-based uncertainty $\mathfrak{U} = \left\{\xi \in \mathbb{R}^k : \|\xi - \widehat{\xi}\| \leq r\right\}$.

4. Polytopic uncertainty $\mathfrak{U} = conv\{\xi^1, \dots \xi^l\}$.

5. Constraint-wise uncertainly $\mathfrak{U} = \mathfrak{U}_1 \times \dots \times \mathfrak{U}_m$, where $\mathcal{U}_i$ affects only the constraint i.

3. Robustness models

This section aims to formally define the main concepts of robustness used to solve optimization problems with uncertainty. In each of the ways to approach robustness, the intuition that exists behind the definition is described; later, the sets that model the uncertainty are characterized and then the problem is written in its robust version. Finally, articles where the definition has been used are referenced.

3.1. Strict robustness

Let $x \; in S$ be a solution to the optimization problem with uncertainty $(P(\xi), \xi \in \mathfrak{U})$. The solution is strict if x is feasible for all possible scenarios of $\mathfrak{U}$, that is, if $F(x, \xi) \leq 0$ for all $\xi \in \mathfrak{U}$. This approach is the most intuitive when trying to solve the optimization problem in a robust way. Formally, consider the set of all possible strictly robust solutions with respect to the uncertainty set $\mathcal{U}$ given by:

$$\mathfrak{F}(\xi) = \{x \in S : F(x, \xi) \leq 0\}$$
$$R(\mathfrak{U}) = \bigcap_{\xi \in \mathfrak{U}} \mathfrak{F}(\xi) \qquad (3)$$

Then the strict robust problem corresponds to the problem formulated in Eq. (4),

$$\min \sup_{\xi \in \mathfrak{U}} f(x, \xi)$$
$$\text{s.t.} \ \ x \in R(\mathfrak{U}) \tag{4}$$
$$x \in S$$

To the best of our knowledge, the first to use strict robustness was Soster in [21], where he applied uncertainty to convex sets, solving the problem using linear programming. Later, this work was extended and placed in a theoretical framework in the articles [22, 23]. The essence of strict robustness is that all scenarios can occur and all of them have an important criticality. In real problems, this type of robustness is necessary in critical systems where a failure is not tolerable. For example, the case of air planes and nuclear plants. However, in other types of problems, such as revenue management, public or scheduling, this type of robustness can be relaxed.

3.2. Cardinality constrained robustness

One way to relax the strict robustness is to restrict the space of uncertainty. There are several ways to achieve this restriction. In cardinality constrained robustness, the property is used that it is unlikely that all the uncertainty parameters change at the same time when analyzing the worst case. Then, we can restrict the cardinality of the uncertainty space by varying only some parameters; the others are modeled with their representative values.

Let $X = \{x_1, \ldots x_n\}$ and $b_1 x_1, \ldots, b_n x_n \leq c$ be a solution and restriction respectively of the optimization problem. Let $\mathfrak{U} = \left\{ b \in \mathbb{R}^n : b_i \in \left[\widehat{b}_i - d_i, \widehat{b}_i + d_i \right], i = 1, \ldots n \right\}$; then, the cardinality constrained robustness is described in Eq. (5).

$$\sum_{i=1}^{n} \widehat{b}_i x_i + \max_{R \subset \{1,\ldots,n\}, |R|=\gamma} \left(\sum_{i \in R} d_i |x_i| \right) \leq c \tag{5}$$

This approach was conceptualized by Bertsimas and Sim [16] for continuous problems. Later, this approach was extended to combinatorial problems in the articles [24, 25].

3.3. Adjustable robustness

Another way to relax the space of uncertainty of strict robustness, corresponds to divide the space into groups of variables. A first group will be called *here and now variables*. These variables correspond to variables that must be evaluated before the scenario $\xi \in \mathfrak{U}$ is determined and the *wait and see variables*, which can be determined once the scenario ξ is known.

Let X be one point of our search space; then, $X = (u, v)$ can be divided into $u \in S^1 \subset \mathbb{R}^{n_1}$ and $v \in S^2 \subset \mathbb{R}^{n_2}$ where $n_1 + n_2 = n$. Then the variables u correspond to the group *here and now variables* and the variables v to the group *wait and see variables*. Formally, this is written in Eq. (6).

$$\min f(u, v, \xi)$$
$$F(u, v, \xi) \leq 0 \tag{6}$$
$$(u, v) \in S^1 \times S^2$$

Then once we have fixed the variables *here and now*, we must make sure that for any of the selected $\xi \in \mathfrak{U}$ scenarios, there is $v \in S^2$ such that (u, v) is feasible for ξ. Let $P_{S^1}(\mathfrak{F}(\xi))$, defined in Eq. (7) be the projection of $\mathfrak{F}(\xi)$ over S^1.

$$P_{S^1}(\mathfrak{F}(\xi)) := \{u \in S^1 : \exists v \in S^2 \text{ s.t. } (u, v) \in \mathfrak{F}(\xi)\} \tag{7}$$

where $\mathfrak{F}(\xi)$ corresponds to the solution space that complies with the constraints defined in Eq. (3). Then, the set of solutions for the split robustness is given by:

$$R = \{u \in S^1 : \forall \xi \in \mathfrak{U} \exists v \in S^2 \text{ s.t. } (u, v) \in \mathfrak{F}(\xi)\}$$
$$= \bigcap_{\xi \in \mathfrak{U}} P_{S^1}(\mathfrak{F}(\xi)) \tag{8}$$

Given a u, the worst case w for some specific u with respect to the set of solutions R, is given by Eq. (9).

$$w^R(u) = \sup_{\xi \in \mathfrak{U}} \inf_{v:(u,v) \in \mathfrak{F}(\xi)} f(u, v, \xi) \tag{9}$$

And therefore, the split robustness is given by Eq. (10).

$$\min\{w^R(u) : u \in R\} \tag{10}$$

The first one to introduce the concept of adjustable robustness was Ben Tal et al. [17] applied to uncertainly problems in linear programming. However, the concept has continued to develop and adapt and nowadays, applications are being seen in portfolio selection [26], in power systems [27], capacity extension planning [28], aperiodic timetabling [29], among others.

3.4. Light robustness

A completely different way of relaxing the concept of strict robustness corresponds to instead of reducing the space of uncertainty, we can relax the constraints in favor of the quality of the solution. This new concept that is called light robustness, this concept considers as a funda-mental hypothesis that if we are able to adequately solve the optimization problem consider-ing the nominal (or average) case, the solution should not be bad and basically, we can concentrate on finding relatively close solutions of the fitness that also fulfill in the best possible way the restrictions of the problem considering all $\xi \in \mathfrak{U}$. Formally, light robustness is detailed in Eq. (11).

$$\min \sum_{i=1}^{k} w_i \lambda_i$$

$$\text{s.t. } f\left(x, \widehat{\xi}\right) \le f^*\left(\widehat{\xi}\right) + \rho \qquad (11)$$

$$F(x, \xi) \le \lambda, \forall \xi \in \mathfrak{U}$$

$$x \in S, \lambda \in \mathbb{R}^k$$

The concept of light robustness was introduced by Fischetti and Monaci [30], the main objective of its new definition was to allow a trade-off between robustness and quality of the solution. A constraint is added by entering the parameter ρ. This parameter forces the solution to have a certain closeness to the solution for the nominal case represented by $\widehat{\xi}$. Because there is a trade-off between quality and robustness, to allow this closeness of the nominal case, it is necessary to relax the original constraints. This is done with the λ factor, where we finally want to find the best set of coefficients that relax our solution.

Originally, the concept of light robustness was conceived to be applied to problems of linear programming [30] and specifically, in time optimizations in Italian single-line instances. Later, in [31] light robustness was applied to determine the best route to traveling in a public transport network in Germany. Later in [18] the concept was generalized taking into account any optimization problem and any set of uncertainty.

3.5. Regret robustness

The regret robustness described by [32] uses a way to relax the problem through the objective function. Let $f^*(\xi)$ be the best target value in the scenario $\xi \in \mathfrak{U}$. Instead of minimizing the worst-case performance of a solution, it minimizes the difference to the objective function of the best solution that would have been possible in a scenario. The regret robustness formulation is shown Eq. (12).

$$\min \sup_{\xi \in \mathfrak{U}} (f(x, \xi) - f^*(\xi))$$

$$\text{s.t. } F(x) \le 0 \qquad (12)$$

$$x \in S$$

Today, we see used in the concept of regret robustness in different areas. In [33], it was used in portfolio optimization problems. In safety investment problems, it was used in [34]. In [35] it was used to solve evacuation planning models.

3.6. Recoverable robustness

Recoverable robustness uses the concept of *recovery algorithm* and, like adjustable robustness, it obtains the solution in two stages. Give a family of algorithms $\mathfrak{A}$. A solution x is recovery robust with respect to $\mathfrak{A}$ if it exists for every scenario $\xi \in \mathfrak{U}$, an algorithm $A \in \mathfrak{A}$ such that A

applied to the solution x and a scenario ξ allows you to build a solution $A(x, \xi) \in \mathfrak{F}(\xi)$. Then the optimization problem in its robust form is written by Eq. (13):

$$\min_{(x,A) \in (\mathfrak{F}(\xi) \times \mathfrak{A})} f(x)$$
$$\text{s.t.} \quad A(x, \xi) \in \mathfrak{F}(\xi), \forall \xi \in \mathfrak{U}$$

(13)

The concept of recoverable robustness was developed in the article [36] applied to shunting problems and later refined in [37] applying recoverable robustness to railway problems with linear programming. Today, we find the concept of recoverable robustness applied to location planning [38], scheduling and delivering routing [39], allocation and network design problems [40], robust traveling salesman problem [41] and transit network design [42], among others.

4. Application areas of robust optimization

In this section, we will describe some examples where robust optimization has been applied. Mainly identified areas have been logistics, finance, water management, energy management and machine learning.

4.1. Energy management

Energy management has received significant attention with respect to robust optimization. In [43] a strategic planning model applied to the integrated oil chain was designed. For the design, it was considered as sources of uncertainty: crude oil production, demand for refined products and market prices. The robust version of the demands for a power plant problem was studied in [44]. In this article the phases of unit commitment and economic dispatch were considered to minimize the local cost. A robust model of energy distribution under uncertainties with respect to wind energy was studied in [45]. In this article, it was shown that the proposed method can be solved in suitable times in addition to being able to effectively capture the ambiguous distribution of wind power generation. In [46], the configuration of the energy consumption of household appliances under the uncertainty of manually operated devices (MOAs) was modeled as a problem of robust optimization. When evaluating all the possible cases of the energy of MOAs, the traditional approach was chosen, that is, using the worst case with the intention of reducing the payment of electricity for all the household appliances. To determine the reduction in the payment, the price of electricity in real time was considered as information in addition to the inclining block rate.

4.2. Water management

In [47] robust optimization was used to handle the uncertainties of water planning resources. In [48] the authors developed a new methodology for the optimizing daily operations of pumping stations. This methodology takes into consideration the fact that a water distribution system is actually unavoidably affected by uncertainties. A multi-objective robust decision-making approach was developed in [49]. This approach supports seasonal water management. In [51], a comparison of Robust Optimization and Info-Gap Methods for Water Resource Management under Deep Uncertainty was made. A multi-objective design of water distribution systems under uncertainty was developed in [50]. The main objectives are (1) minimize the total water distribution system (WDS) design cost and (2) maximize WDS robustness. In the article, the WDS robustness is defined as the probability of simultaneously satisfying minimum pressure head constraints at all nodes in the network.

4.3. Machine learning

In [52] regularized support vector machines (SVMs) were considered, and they were shown to be equivalent to a robust formulation of the problem. The authors show that this equivalence between robust optimization and regularization has implications for both algorithms and analysis. The equivalence of robustness and regularization provides a robust optimization interpretation for the success of regularized SVMs. On the other hand, Fertis in his doctoral thesis [53], studied the connection between regularizations like Lazo and robustness. Specifically, he showed that in classical regression, regularized estimators like lasso can be obtained by applying robust optimization to the classical least squares problem. He discovers an explicit connection between the size and the structure of the uncertainty used in the robust estimator, with the coefficient and the kind of norm used in regularization. Xu et al. [54], investigated a probabilistic interpretation of robust optimization. They established a connection between robust optimization and distributionally robust stochastic programming (DRSP). In the article, they showed that the solution to any robust optimization problem is also a solution to a DRSP problem. In [55] the problem of constructing robust classifiers when the training is subject to uncertainty was studied. The problem is posed by a chance-constrained programming, which ensures that the uncertainty of the data is correctly defined with high probability.

4.4. Logistics

In the area of logistics problems such as the traveling salesman and routing problem have been explored in their robust versions. A Swarm intelligence system was designed in [56] to solve the vehicle routing problem with time windows and uncertain travel times. The uncertainty here models the perturbation in the data. This perturbation, is caused by the effects of unpredictable events, such as traffic jams, road building, etc. In the article, the authors proposed a heuristic approach using ant colony optimization as a metaheuristic. In [57], the open vehicle routing problem with uncertain demands was studied. In this problem, the vehicles have as an additional function that they do not necessarily return to their original locations after delivering the products to the customers. First, the authors modeled the demand of the clients as specific sets of limited uncertainty with expected values of demand and nominal values. Having the sets modeled, they later proposed a robust optimization model that aims to minimize transport costs and unsatisfied demands on the specific uncertainty sets defined. The robust vehicle routing problem with time windows was solved in [58]. They proposed two new formulations for the robust problem, each based on a different robust approach. They proposed two new formulations for the robust problem, each based on a different robust approach. The first formulation uses adjustable robustness with the aim of extending the well-known formulation of resource inequalities. The second formulation generalizes a path inequalities formulation to the uncertain context. In this case, uncertainty is modeled in the formulation of the problem. In [41], an uncertain traveling salesman problem was developed. In this problem, the distances between the nodes are not exactly known, but they can be obtained from a set of uncertainties of possible scenarios. This set of uncertainties is modeled as intervals, including an additional limit associated with the number of distances that can deviate from their expected nominal values. In the study, a recoverable robust model was proposed. This model allows a tour to change only a limited number of borders once a scenario is known; all these with the goal of minimizing the complexity in calculations. The robust traveling salesman problem with interval data was studied in [59]. In the article, travel times are specified as a range of possible values. They applied the robust deviation criterion to drive the optimization over the interval data problem thus obtained.

Another interesting group of problems in the logistics area corresponds to facility location problems. The robust formulation of these problems aims to obtain an optimal design of a system considering uncertainty. The authors introduce a robust optimization-based approach to obtain some capacity expansion solutions that are not sensitive to this uncertainty. In this area, we highlight the work carried out by [60], where they considered the question of how to make a decision about capacity expansions for a network flow problem that is subject to demand and travel time uncertainty. The authors introduce a robust optimization-based approach to obtain some capacity expansion solutions that are not sensitive to this uncertainty. They show that the robust modeled solution is a computationally tractable problem when considering general uncertainty sets together with reasonable conditions for network flow applications. Another interesting problem in this area is the robust transmission expansion planning. In [61] the authors address the problem of transmission expansion planning, considering uncertainties in the electric power system. They consider varied sources of uncertainty such as: the growth of future demand, the availability of generation facilities, geographical characterization within the electric power system. A robust adaptive optimization model is used to obtain investment decisions with the objective of minimizing the total costs of the system and anticipating the worst-case materialization of the uncertain parameters within a uncertainty set.

4.5. Public goods

Public goods can be understood as a merchandise or service that is provided non-profit to all members of a society. This merchandise or service, can be provided by the government, an individual or an organization. When we consider public goods and robust optimization, interesting applications appear. An interesting first application corresponds to radiation therapy. When a radiation therapy examination is performed, there are uncertainties that are fundamental to consider in defining the correct treatment in patients with cancer. In this context, addressing problems through robust optimization makes a lot of sense. In [62] the authors constructed an uncertainty model of the movement of respiration based on probability density functions. These functions allowed them to robustly model the optimization of intensity-modulated radiation therapy.

Another interesting implementation associated with the application of robust optimization to public goods corresponds to intrahospital transport. Intrahospital transport is often required for reasons associated with a diagnosis or some therapy that the patient must perform. Depending on the design of the hospital, transportation between the nursing rooms and the service units is provided by ambulances or by trained personnel accompanying patients on foot. When the hospital is large, the patient transport service is often poorly managed and there is no associated flow coordination; on the other hand, there is no clarity of all the necessary transports since they are dependent on the diagnoses. In [63] the authors address the problem of defining robustness to patient flow management in the context of optimized patient transport in hospitals. In [64] a methodology was proposed to obtain a robust logistics plan to mitigate the uncertainty of the demand for humanitarian relief supply chains. More specifically, the authors formulated the problem as a robust optimization problem with the objective of dynamically assigning emergency response and generate evacuation traffic flow, all this in the context of time-dependent demand uncertainty.

5. Discussion and conclusions

In this article, we have carefully reviewed the different definitions that have appeared in the literature to address the concept of robust optimization. We have taken special care to formalize each of the definitions and cite specific examples where they have been used. Subsequently, a review was made in areas where robust optimization has been applied. In particular, the areas of water management, energy management, machine learning, logistics and public goods stood out. With the advent of the concepts and technologies associated with the Internet of Things and Big Data, it is expected that the problems described above have a greater amount of data to build more robust models; however, this brings challenges regarding the complexity of the algorithms, in addition to the learning and operation of these in real time.

When we analyze the research works developed in the area of robust optimization, we found that there is a lack of a formal argument that clearly defines the uncertainty set to be used to solve the problem in a robust way. Usually, the choice is guided by business intuition together with the need to adapt the uncertainty set to solve the problem in a reasonable time.

Therefore, there is an important space to develop quantitative studies to determine what kind of robustness and uncertainty set should be used to solve a problem. Identifying how different uncertainty sets behave for a defined problem is fundamental. To be able to answer questions such as: How is the quality of the solutions perturbed with the choice of the uncertainty set?, Is this perturbation important for the problem that is being solved?, How is the convergence of the algorithm altered against different sets of uncertainty?, Can we classify problems according to some degree of robustness? Can this classification be related to the type of uncertainty to be used? The answer to these questions allows developing a methodology that allows identifying which is the robustness required by the problem, what type of uncertainty set should be chosen and how is the behavior of the algorithm in terms of quality of its results and convergence.

As future lines of research in the area of robust optimization, we see that considering these group of definitions together with the different applications mentioned earlier, we can work on developing a methodology that gives a specific problem, allows in a simple way to identify which definition is the most appropriate and which methods they are the most appropriate to solve the problem at reasonable times.

Regarding the tractability of robust problems, we have not found solutions where the hybridization of metaheuristics with other techniques is exploited such as integration with mathematical programming, with simulations or integration with machine learning, all these with the goal of improving convergence times of algorithms.

Particularly, according to our experience in the integration of machine learning and metaheuristics, a line that must be explored corresponds to the use of a general scheme of integration of these two areas through the use of metalearning techniques. Considering that we have a set of algorithms or settings of some algorithm, we use a mechanism that selects the best algorithm or settings for given an instance to obtain the best convergence and results. Furthermore, the use of reinforced learning can be explored to enrich the metamodel with the new results generated.

Author details

José García[1,2]* and Alvaro Peña[1]

*Address all correspondence to: joseantonio.garcia@telefonica.com

1 Escuela de Ingeniería en Construcción, Pontificia Universidad Católica de Valparaíso, Valparaíso, Chile

2 Telefónica I+D, Providencia, Santiago, Chile

References

[1] Graells-Garrido E, García J. Visual exploration of urban dynamics using mobile data. In: International Conference on Ubiquitous Computing and Ambient Intelligence. Springer; 2015. pp. 480-491

[2] Graells-Garrido E, Peredo O, García J. Sensing urban patterns with antenna mappings: The case of Santiago, Chile. Sensors. 2016;**16**(7):1098

[3] Peredo OF, García JA, Stuven R, Ortiz JM. Urban dynamic estimation using mobile phone logs and locally varying anisotropy. In: Geostatistics Valencia 2016; Springer; 2017. pp. 949-964

[4] García J, Pope C, Altimiras F. A distributed k-means segmentation algorithm applied to lobesia botrana recognition. Complexity. 2017;**2017**

[5] García J, Crawford B, Soto R, García P. A multi dynamic binary black hole algorithm applied to set covering problem. In: International Conference on Harmony Search Algorithm. Singapore: Springer; 2017. pp. 42-51

[6] Crawford B, Soto R, Monfroy E, Astorga G, García J, Cortes E. A meta-optimization approach for covering problems in facility location. In: Workshop on Engineering Applications. Vol. 742. 2018. pp. 565-578

[7] García J, Crawford B, Soto R, Castro C, Paredes F. A k-means binarization framework applied to multidimensional knapsack problem. Applied Intelligence. Springer; 2018;**48** (2):357-380

[8] García J, Crawford B, Soto R, Astorga G. A percentile transition ranking algorithm applied to knapsack problem. In: Proceedings of the Computational Methods in Systems and Software. Springer; 2017. pp. 126-138

[9] Rooderkerk RP, van Heerde HJ. Robust optimization of the 0–1 knapsack problem: Balancing risk and return in assortment optimization. European Journal of Operational Research. 2016;**250**(3):842-854

[10] Jin Y, Branke J. Evolutionary optimization in uncertain environments-a survey. IEEE Transactions on Evolutionary Computation. 2005;**9**(3):303-317

[11] Paenke I, Branke J, Jin Y. Efficient search for robust solutions by means of evolutionary algorithms and fitness approximation. IEEE Transactions on Evolutionary Computation. 2006;**10**(4):405-420

[12] Gabrel V, Murat C, Thiele A. Recent advances in robust optimization: An overview. European Journal of Operational Research. 2014;**235**(3):471-483

[13] Jin Y, Sendhoff B. Trade-off between performance and robustness: An evolutionary multiobjective approach. In: EMO. Vol. 3. Springer; 2003. pp. 237-251

[14] Lim D, Ong Y-S, Lim M-H, Jin Y. Single/multi-objective inverse robust evolutionary design methodology in the presence of uncertainty. In: Evolutionary Computation in Dynamic and Uncertain Environments. Springer; 2007. pp. 437-456

[15] Ben-Tal A, Ghaoui L.E., Nemirovski A. Robust Optimization. Princeton Series in Applied Mathematics. Princeton University Press; 2009. ISBN: 9781400831050. https://books.google.cl/books?id=DttjR7IpjUEC

[16] Bertsimas D, Sim M. The price of robustness. Operations Research. 2004;**52**(1):35-53

[17] Ben-Tal A, Goryashko A, Guslitzer E, Nemirovski A. Adjustable robust solutions of uncertain linear programs. Mathematical Programming. 2004;**99**(2):351-376

[18] Schöbel A. Generalized light robustness and the trade-off between robustness and nominal quality. Mathematical Methods of Operations Research. 2014;**80**(2):161-191

[19] Ben-Tal A, Bertsimas D, Brown DB. A soft robust model for optimization under ambiguity. Operations Research. 2010;**58**(4-part-2):1220-1234

[20] Jenkins L. Selecting scenarios for environmental disaster planning. European Journal of Operational Research. 2000;**121**(2):275-286

[21] Soyster AL. Convex programming with set-inclusive constraints and applications to inexact linear programming. Operations Research. 1973;**21**(5):1154-1157

[22] Ben-Tal A, Nemirovski A. Robust convex optimization. Mathematics of Operations Research. 1998;**23**(4):769-805

[23] Ben-Tal A, Nemirovski A. Robust solutions of uncertain linear programs. Operations Research Letters. 1999;**25**(1):1-13

[24] Atamtürk A. Strong formulations of robust mixed 0–1 programming. Mathematical Programming. 2006;**108**(2):235-250

[25] Goetzmann K-S, Stiller S, Telha C. Optimization over integers with robustness in cost and few constraints. In: WAOA. Vol. 2011. Springer; 2011. pp. 89-101

[26] Fliedner T, Liesiö J. Adjustable robustness for multi-attribute project portfolio selection. European Journal of Operational Research. 2016;**252**(3):931-946

[27] Ding T, Bie Z, Bai L, Li F. Adjustable robust optimal power flow with the price of robustness for large-scale power systems. IET Generation, Transmission & Distribution. 2016;**10**(1):164-174

[28] Mejia-Giraldo D, McCalley J. Adjustable decisions for reducing the price of robustness of capacity expansion planning. IEEE Transactions on Power Systems. 2014;**29**(4):1573-1582

[29] Goerigk M, Schöbel A. Recovery-to-optimality: A new two-stage approach to robustness with an application to aperiodic timetabling. Computers & Operations Research. 2014;**52**:1-15

[30] Fischetti M, Monaci M. Light robustness. In: Robust and Online Large-Scale Optimization. Springer; 2009. pp. 61-84

[31] Goerigk M, Schmidt M, Schöbel A, Knoth M, Müller-Hannemann M. The price of strict and light robustness in timetable information. Transportation Science. 2013;**48**(2):225-242

[32] Kouvelis P, Yu G. Robust Discrete Optimization and Its Applications. Vol. 14. US: Springer Science & Business Media; 2013

[33] Xidonas P, Mavrotas G, Hassapis C, Zopounidis C. Robust multiobjective portfolio optimization: A minimax regret approach. European Journal of Operational Research. 2017;**262**(1):299-305

[34] Aven T, Hiriart Y. Robust optimization in relation to a basic safety investment model with imprecise probabilities. Safety Science. 2013;**55**:188-194

[35] Goerigk M, Hamacher HW, Kinscherff A. Ranking robustness and its application to evacuation planning. European Journal of Operational Research. Elsevier; 2018;**264**(3):837-846

[36] Cicerone S, D'Angelo G, Di Stefano G, Frigioni D, Navarra A. 12. Robust algorithms and price of robustness in shunting problems. In: Liebchen C, Ahuja KR, Mesa AJ, editor. 7th Workshop on Algorithmic Approaches for Transportation Modeling, Optimization, and Systems (ATMOS'07). Vol. 7. Dagstuhl, Germany: Schloss Dagstuhl-Leibniz-Zentrum für Informatik; 2007. ISBN: 978-3-939897-04-0. ISSN: 2190-6807. DOI: 10.4230/OASIcs.ATM OS.2007.1175

[37] Liebchen C, Lübbecke M, Möhring R, Stiller S. The concept of recoverable robustness, linear programming recovery, and railway applications. In: Robust and Online Large-Scale Optimization. Berlin, Heidelberg: Springer; 2009. pp. 1-27

[38] Carrizosa E, Goerigk M, Schöbel A. A biobjective approach to recoverable robustness based on location planning. European Journal of Operational Research. 2017;**261**(2):421-435

[39] Cheref A, Artigues C, Billaut J-C. A new robust approach for a production scheduling and delivery routing problem. IFAC-PapersOnLine. 2016;**49**(12):886-891

[40] Kutschka M. Robustness concepts for knapsack and network design problems under data uncertainty. In: Operations Research Proceedings 2014. Cham: Springer; 2016. pp. 341-347

[41] Chassein A, Goerigk M. On the recoverable robust traveling salesman problem. Optimization Letters. 2016;**10**(7):1479-1492

[42] Cadarso L, Marn Á. Rapid transit network design considering risk aversion. Electronic Notes in Discrete Mathematics. 2016;**52**:29-36

[43] Ribas GP, Hamacher S, Street A. Optimization under uncertainty of the integrated oil supply chain using stochastic and robust programming. International Transactions in Operational Research. 2010;**17**(6):777-796

[44] Zhang M, Guan Y. Two-Stage Robust Unit Commitment Problem. USA: University of Florida; 2009

[45] Xiong P, Singh C. Distributionally robust optimization for energy and reserve toward a low-carbon electricity market. Electric Power Systems Research. 2017;**149**:137-145

[46] Du Y, Jiang L, Li Y, Wu Q. A robust optimization approach for demand side scheduling considering uncertainty of manually operated appliances. IEEE Transactions on Smart Grid. 2018;**9**(2):743-755. ISSN: 1949-3053. DOI: 10.1109/TSG.2016.2564159

[47] Beh EH, Zheng F, Dandy GC, Maier HR, Kapelan Z. Robust optimization of water infrastructure planning under deep uncertainty using metamodels. Environmental Modelling & Software. 2017;**93**:92-105

[48] Goryashko AP, Nemirovski AS. Robust energy cost optimization of water distribution system with uncertain demand. Automation and Remote Control. 2014;**75**(10):1754-1769

[49] Riegels N, Jessen O, Madsen H. Using multi-objective robust decision making to support seasonal water management in the chao phraya river basin, Thailand. In: EGU General Assembly Conference Abstracts. Vol. 18. 2016. p. 12712

[50] Roach T, Kapelan Z, Ledbetter R, Ledbetter M. Comparison of robust optimization and info-gap methods for water resource management under deep uncertainty. Journal of Water Resources Planning and Management. 2016;**142**(9):04016028

[51] Kapelan ZS, Savic DA, Walters GA. Multiobjective design of water distribution systems under uncertainty. Water Resources Research. 2005;**41**(11)

[52] Xu H, Caramanis C, Mannor S. Robustness and regularization of support vector machines. Journal of Machine Learning Research. 2009;**10**(Jul):1485-1510

[53] Fertis A. A robust optimization approach to statistical estimation problems by Apostolos G. Fertis [PhD thesis]. Massachusetts Institute of Technology; 2009

[54] Xu H, Caramanis C, Mannor S. A distributional interpretation of robust optimization. Mathematics of Operations Research. 2012;**37**(1):95-110

[55] Ben-Tal A, Bhadra S, Bhattacharyya C, Nath JS. Chance constrained uncertain classification via robust optimization. Mathematical Programming. 2011;**127**(1):145-173

[56] Toklu NE, Gambardella LM, Montemanni R. A multiple ant colony system for a vehicle routing problem with time windows and uncertain travel times. Journal of Traffic and Logistics Engineering. 2014;**2**(1)

[57] Cao E, Lai M, Yang H. Open vehicle routing problem with demand uncertainty and its robust strategies. Expert Systems with Applications. 2014;**41**(7):3569-3575

[58] Agra A, Christiansen M, Figueiredo R, Hvattum LM, Poss M, Requejo C. The robust vehicle routing problem with time windows. Computers & Operations Research. 2013;**40**(3):856-866

[59] Montemanni R, Barta J, Mastrolilli M, Gambardella LM. The robust traveling salesman problem with interval data. Transportation Science. 2007;**41**(3):366-381

[60] Ordóñez F, Zhao J. Robust capacity expansion of network flows. Networks. 2007;**50**(2):136-145

[61] Ruiz C, Conejo AJ. Robust transmission expansion planning. European Journal of Operational Research. 2015;**242**(2):390-401

[62] Bortfeld T, Chan TC, Trofimov A, Tsitsiklis JN. Robust management of motion uncertainty in intensity-modulated radiation therapy. Operations Research. 2008;**56**(6):1461-1473

[63] Hanne T, Melo T, Nickel S. Bringing robustness to patient flow management through optimized patient transports in hospitals. Interfaces. 2009;**39**(3):241-255

[64] Ben-Tal A, Do Chung B, Mandala SR, Yao T. Robust optimization for emergency logistics planning: Risk mitigation in humanitarian relief supply chains. Transportation Research Part B: Methodological. 2011;**45**(8):1177-1189

Evaluation of Non-Parametric Selection Mechanisms in Evolutionary Computation: A Case Study for the Machine Scheduling Problem

Maxim A. Dulebenets

Additional information is available at the end of the chapter

http://dx.doi.org/10.5772/intechopen.75984

Abstract

Evolutionary Algorithms have been extensively used for solving stochastic, robust, and dynamic optimization problems of a high complexity. Selection mechanisms play a very important role in design of Evolutionary Algorithms, as they allow identifying the parent chromosomes, that will be used for producing the offspring, and the offspring chromosomes, that will survive in the given generation and move on to the next generation. Selection mechanisms, reported in the literature, can be classified in two groups: (1) parametric selection mechanisms, and (2) non-parametric selection mechanisms. Unlike parametric selection mechanisms, non-parametric selection mechanisms do not have any parameters that have to be set, which significantly facilitates the Evolutionary Algorithm parameter tuning analysis. This study presents a comprehensive analysis of the commonly used non-parametric selection mechanisms. Comparison of the selection mechanisms is performed for the machine scheduling problem. The objective of the presented mathematical model is to determine the assignment of the arriving jobs among the available machines, and the processing order of jobs on each machine, aiming to minimize the total job processing cost. Different categories of Evolutionary Algorithms, which deploy various non-parametric selection mechanisms, are evaluated in terms of the objective function value at termination, computational time, and changes in the population diversity. Findings indicate that the Roulette Wheel Selection and Uniform Sampling selection mechanisms generally yield higher population diversity, while the Stochastic Universal Sampling selection mechanism outperforms the other non-parametric selection mechanisms in terms of the solution quality.

Keywords: optimization, Evolutionary Algorithms, non-parametric selection mechanisms, machine scheduling problems, parameter tuning, computational time

1. Introduction

Evolutionary Algorithms (EAs) and other metaheuristic algorithms have been widely used for solving complex stochastic, robust, and dynamic optimization problems. These complex problems include but are not limited to the following: vertex cover problem, Boolean satisfiability problem, maximum clique size problem, Knapsack problem, traveling salesman problem, bin packing problem, machine scheduling problems, and others [1, 2]. Some of the aforementioned problems have a non-deterministic polynomial time complete (NP-complete) complexity, while the others are non-deterministic polynomial time hard (NP-hard). The exact solution algorithms cannot be used to solve NP-complete and NP-hard problems to the global optimality for the realistic size problem instances within an acceptable computational time. On the other hand, the approximation algorithms, including EAs and other metaheuristic algorithms, are able to provide good quality solutions within a reasonable computational time. Candidate solutions to the problem of interest are encoded in the chromosomes within EAs. Different types of chromosome representations have been reported in the EA literature. For example, canonical Genetic Algorithms, developed by Holland, rely on a binary chromosome representation; while canonical Evolutionary Strategies, proposed by Rechenberg, use a real-valued chromosome representation [3, 4]. On the other hand, Genetic Programming, developed by Koza, relies on a tree-based chromosome representation [3, 4].

Once the chromosome representation is selected, the initial population is generated, and fitness values of the initial population chromosomes are estimated. Then, the EA starts an iterative process, where the population chromosomes are continuously altered using selection and EA operators (e.g., crossover and mutation) from one generation to another, aiming to identify superior solutions. The EA is terminated, once a certain stopping criterion is met (in some EAs multiple stopping criteria can be imposed). Two types of selection mechanisms are applied throughout the EA evolution: (1) parent selection, which aims to identify a subset of individuals from the offspring chromosomes, survived in the previous generation, that will participate in the EA operations and generate the new offspring chromosomes; and (2) offspring selection, which aims to identify a subset of individuals from the generated offspring chromosomes that will survive in the given generation and will be moved to the next generation. A large number of different selection mechanisms have been reported in the EA literature, which can be categorized in two groups: (1) parametric selection mechanisms (e.g., Exponential Ranking Selection, Tournament Selection, Boltzmann Selection), and (2) non-parametric selection mechanisms (e.g., Roulette Wheel Selection, Stochastic Universal Sampling, Binary Tournament Selection, Ranking Selection, Uniform Sampling).

Each EA has several parameters (e.g., population size, crossover probability, mutation probability, and others), which are generally determined based on a parameter tuning [3, 4]. A "full factorial design" methodology has been widely used for the EA parameter tuning [5]. Based on the latter methodology, the algorithm has a number of parameters (or factors - f), which have a set of candidate values (or levels - l). In order to set the appropriate EA parameter values, a total of l^f algorithmic runs will be required throughout the parameter tuning analysis. Based on the analysis of a tradeoff between the objective function and computational time values, the most promising parameter combination will be chosen. Parametric selection mechanisms will increase the number of algorithmic runs to $l^{(f+N^{SEL})}$, where N^{SEL} – is the number of parameters for a given selection mechanism. Such increase in the number of algorithmic runs can make the parameter tuning analysis computationally prohibitive due to significant computational time required. Moreover, the parameter values of the selection mechanisms, adopted for a given set of problem instances, may worsen the EA performance, when applied to a different set of problem instances.

In order to avoid the latter drawbacks and facilitate the EA parameter tuning analysis, this study solely applies non-parametric selection mechanisms throughout the EA design. Different EA categories, which rely on various non-parametric selection mechanisms, are evaluated based on the major algorithmic performance indicators, including the objective function value at termination, computational time, and changes in the population diversity throughout the algorithmic evolution. The computational experiments are conducted for the machine scheduling problem. The machine scheduling problem deals with allocation of the available handling resources (i.e., machines) for service of the tasks (i.e., jobs), which arrive at the given facility with a specific frequency [2]. The machine scheduling problem receives an increasing attention from the community, as it is considered as an important decision problem in manufacturing, service industries, and supply chain management [6–10]. Without efficient sequencing and scheduling, the supply chain players may not be able to meet specific deadlines, which are established for processing certain products. The latter may incur substantial monetary losses and, ultimately, can even result in the customer loss. In the meantime, poor utilization of the available handling resources may cause drastic monetary losses as well. Therefore, development of advanced decision support tools for the machine scheduling problems (including effective solution algorithms, which are the primary focus of this study) becomes critical in the current competitive environment.

Findings from this research are expected to provide important insights regarding non-parametric selection mechanisms, which can be further used in future for the design of EAs. Efficient non-parametric selection mechanisms will be critical for Hybrid EAs, which along with the standard EA parameters (e.g., population size, crossover probability, mutation probability) may require setting additional parameters for the local search heuristics. The remaining sections of this chapter are organized in the following order. The next section discusses the machine scheduling environment, where the developed EA will be applied. The third section presents a mixed integer mathematical model for the machine scheduling problem. The fourth section focuses on a detailed description of the main EA components. The fifth section discusses the computational experiments, which were conducted in this study for evaluation of non-parametric selection mechanisms. The last section summarizes findings and outlines potential directions for the future research.

2. Machine scheduling

The objective of the machine scheduling problems (MSPs) is to allocate the arriving jobs among the available machines and identify the processing order of jobs on each machine. A large number of various MSPs have been widely studied in the past, such as single machine, identical machines in parallel, machines in parallel with different speeds, unrelated machines in parallel, job shop, and others [2]. The aforementioned MSPs differ in terms of machine properties (e.g., machines at a given facility have identical properties vs. machines at a given facility have different properties), job type (e.g., the processing time of a given job may vary on two machines with the same speeds based on the job type), order of machines to be visited (e.g., a given job may have to be processed on several machines in a specific order), etc.

The unrelated MSP will be studied in this chapter. Let $I = \{1, ..., m\}$ be a set of jobs, arriving at the facility, which should be processed on the available machines within a given planning horizon. Let $J = \{1, ..., n\}$ be a set of machines available at the given facility within a given planning horizon. Let $K = \{1, ..., p\}$ be a set of job processing orders. Each job should be assigned for processing on one of the available machines in one of the processing orders. The machines at the given facility are assumed to have different properties (e.g., different speeds); therefore, the processing time of a given job may vary depending on the machine assignment. Furthermore, the processing time on a given machine depends on the job type (i.e., the processing time for a given job on the machines with the same speed may be different due to the job type). The latter three aspects are common for the unrelated MSPs. The MSP environment, modeled in this study, is illustrated in **Figure 1**.

Once the job arrives at the facility, it will be directed to the assigned machine for processing. If the assigned machine is processing another job at the moment, the arriving job will be queued, while waiting to be processed (see **Figure 1**). It is assumed that the facility operator will incur the job waiting cost ($c_i^{WC}, i \in I$ in USD/hour), as increasing number of waiting jobs may cause congestion at the given facility. Furthermore, the facility operator will incur the cost of processing a given job on one of the available machines ($c_i^{HC}, i \in I$ in USD/hour). Each job, arriving at the facility, must be processed by specific time ($DP_i, i \in I$ in hours). If the job processing deadline is violated, the facility operator will incur the cost due to job processing delays ($c_i^{DC}, i \in I$ in USD/hour). The objective of the facility operator is to allocate the arriving jobs among the available machines and identify the processing order of jobs on each machine,

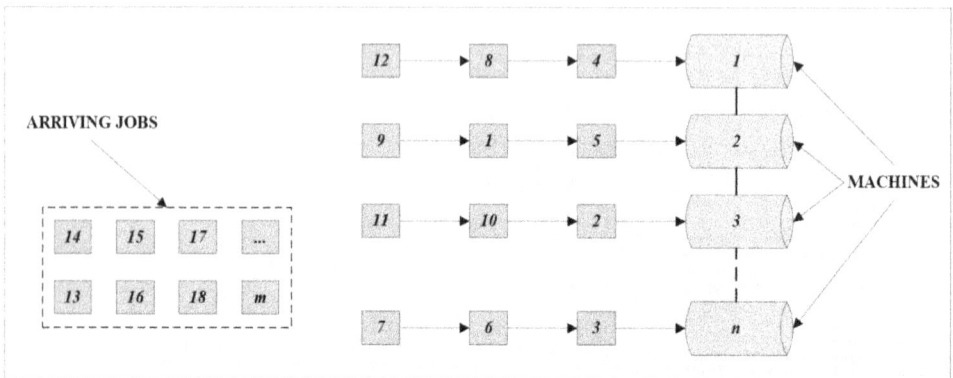

Figure 1. Machine scheduling environment.

aiming to minimize the total job processing cost, which includes: (1) the total job handling cost; (2) the total job waiting cost; and (3) the total cost due to job processing delays.

3. Mathematical model

This section of the chapter presents a mixed integer programming model for the machine scheduling problem (**MSP**), which is studied herein. A detailed description of notations used in the mathematical model and throughout this book chapter is provided at the end of the book chapter.

MSP: Machine Scheduling Problem

$$\min \left[\sum_{i \in I} \sum_{j \in J} \sum_{k \in K} \left(HT_{ij} x_{ijk} c_i^{HC} \right) + \sum_{i \in I} \left(WT_i c_i^{WC} \right) + \sum_{i \in I} \left(PD_i c_i^{DC} \right) \right] \qquad (1)$$

Subject to:

$$\sum_{j \in J} \sum_{k \in K} x_{ijk} = 1 \forall i \in I \qquad (2)$$

$$\sum_{i \in I} x_{ijk} \leq 1 \forall j \in J, k \in K \qquad (3)$$

$$\sum_{i^* \in I: i^* \neq i} \sum_{k^* \in K: k^* < k} \left(HT_{i^* j} x_{i^* jk^*} + IT_{i^* jk^*} \right) + IT_{ijk} - AT_i x_{ijk} \geq 0 \;\; \forall i \in I, j \in J, k \in K \qquad (4)$$

$$SPT_i \geq \sum_{i^* \in I: i^* \neq i} \sum_{k^* \in K: k^* < k} \left(HT_{i^* j} x_{i^* jk^*} + IT_{i^* jk^*} \right) + IT_{ijk} - PN \left(1 - x_{ijk} \right) \forall i \in I, j \in J, k \in K \qquad (5)$$

$$WT_i \geq SPT_i - AT_i \forall i \in I \qquad (6)$$

$$FPT_i \geq SPT_i + \sum_{j \in J} \sum_{k \in K} \left(HT_{ij} x_{ijk} \right) \forall i \in I \qquad (7)$$

$$PD_i \geq FPT_i - DP_i \forall i \in I \qquad (8)$$

The objective function (1) of the **MSP** mathematical model minimizes the total job processing cost, which is composed of the following components: (1) the total job handling cost; (2) the total job waiting cost; and (3) the total cost due to job processing delays. Constraint set (2) guarantees that each job will be scheduled for processing on one of the available machines in one of the processing orders. Constraint set (3) ensures that no more than one job can be processed on each machine in a given processing order. Constraint set (4) ensures that the processing of a given job will not start before its arrival at the facility. Constraint set (5) calculates the start processing time for each job, arriving at the facility. Constraint set (6) computes the waiting time for each job, arriving at the facility. Constraint set (7) estimates the finish processing time for each job. Constraint set (8) calculates hours of delay in processing each job, arriving at the facility.

4. Evolutionary Algorithm description

MSPs belong to the class of NP-hard problems, which cannot be solved using the exact optimization algorithms to the global optimality for the realistic size problem instances within an acceptable computational time. Therefore, a set of EAs were developed in this study to solve the MSP mathematical model. EAs were differentiated based on the type of non-parametric selection mechanism adopted. This section provides an outline of the main EA steps and a detailed description of each step.

4.1. Main EA steps

The main EA steps are presented in **Algorithm 1**. The data structures for the EA variables are initialized in step 0. The initial population is generated in steps 1–2. After that, fitness of the initial population chromosomes is evaluated in step 3. Then, the EA algorithm starts an iterative process (steps 4–12), where the fittest individual is stored before applying the parent selection in step 6. The latter strategy is commonly referred to as "Elitist Strategy" in Evolutionary Computation. After that, the parent chromosomes are determined in step 7, while the offspring chromosomes are produced via the EA operations in step 8. Fitness of the offspring chromosomes is evaluated in step 9. After that, the offspring selection is executed to determine the offspring chromosomes that along with the fittest individual will be moved to the next generation (steps 10 and 11). The iterative process is continuously executed until a certain stopping criterion is met. At convergence, the proposed EA algorithm returns the best solution, which corresponds to the job to machine to processing order assignment with the least possible job processing cost. A detailed description of each EA component is presented in Sections 4.2–4.8.

Algorithm 1. Evolutionary Algorithm (EA).

$EA(Data, \Omega, \sigma^C, \sigma^M, SC)$.

in: *Data* - input data for the MSP mathematical model; Ω - population size; σ^C - crossover probability; σ^M - mutation probability; *SC* - stopping criterion

out: *Solution* - the best job to machine to processing order assignment

0: $|Population| \leftarrow \Omega; |Fitness| \leftarrow \Omega; |Parents| \leftarrow \Omega; |Offspring| \leftarrow \Omega; |Best| \leftarrow \emptyset$

1: $gen \leftarrow 1$

2: $Population_{gen} \leftarrow \textbf{InitPopulation}(Data, \Omega)$

3: $Fitness_{gen} \leftarrow \textbf{FitnessEval}\left(Data, Population_{gen}\right)$

4: **while** $SC \leftarrow FALSE$ **do**

5:    $gen \leftarrow gen + 1$

6:    $Best \leftarrow argmin\left(Fitness_{gen}\right)$

7:    $Parents_{gen} \leftarrow \textbf{ParentSel}\left(Population_{gen}, Fitness_{gen}\right)$

Algorithm 1. Evolutionary Algorithm (EA).

8:     $Offspring_{gen} \leftarrow EAoperation(Parents_{gen}, \sigma^C, \sigma^M)$

9:     $Fitness_{gen} \leftarrow FitnessEval\left(Data, Offspring_{gen}\right)$

10:    $Population_{gen+1} \leftarrow Population_{gen+1} \cup \{Best\}$

11:    $Population_{gen+1} \leftarrow OffspringSel\left(Offspring_{gen}, Fitness_{gen}\right)$

12: **end while**

13: $Solution \leftarrow argmin\left(Fitness_{gen} \cup Fitness_{Best}\right)$

14: **return** $Solution$

4.2. Chromosome representation

Two-dimensional integer chromosomes will be used in this study to represent candidate solutions to the MSP mathematical model (i.e., job to machine to processing order assignments). Note that the term "chromosome" is used interchangeably with the term "individual" throughout this chapter, as both terms represent the same meaning [3]. An example of a chromosome is illustrated in **Figure 2**, where 9 jobs are scheduled for processing on 3 machines. Specifically, jobs "2", "3", and "5" are scheduled for processing on machine "1" (in that specific processing order); jobs "4", "6", and "9" are scheduled for processing on machine "2" (in that specific processing order); while jobs "1", "7", and "8" are scheduled for processing on machine "3" (in that specific processing order). The term "genes" will be used in this study to denote components of a chromosome (i.e., machine identifiers and job identifiers).

4.3. Initialization of the chromosomes and population

There are two major approaches for initializing the chromosomes and population within EAs. The first approach initializes the chromosomes and population randomly (i.e., the job to machine to processing order assignment is determined randomly). The second approach relies on application of the local search heuristics. A large number of the local search heuristics have been presented in the machine scheduling literature, such as [2]: First In First Out, First In Last Out, Shortest Processing Time First, Shortest Remaining Processing Time on the Fastest Machine, Shortest Setup Time First, and others. The local search heuristics may allow obtaining better quality solutions as compared to the random initialization mechanisms. However, the local search heuristics, which have been used for MSPs, are typically deterministic. Therefore, the population, initialized using deterministic local search heuristics, will have identical chromosomes, which will negatively affect the population diversity (i.e., only one domain of the

Machine ID →	1	1	1	2	2	2	3	3	3
Job ID →	2	3	5	4	6	9	1	7	8

Figure 2. Chromosome representation example.

search space will be explored at the population initialization stage). To avoid the latter draw-back and ensure the population diversity, this study will use a random initialization mechanism to create the initial population. The number of individuals in the population is determined based on the population size parameter (Ω).

4.4. Fitness function

The fitness function of chromosomes is assumed to be equal to the objective function of the MSP mathematical model (i.e., total job processing cost). Application of various scaling mechanisms for the fitness function (e.g., linear scaling, sigma truncation, and power law scaling) to control the selection pressure throughout the algorithmic run will be one of the future research directions of this study.

4.5. Parent selection mechanism

The purpose of the parent selection mechanism is to determine a subset of individuals from the offspring chromosomes, survived in the previous generation, that will participate in the EA operations and generate the new offspring chromosomes. As discussed in the introduction section of this chapter, the main objective of this study is to evaluate various non-parametric selection mechanisms, commonly used in the literature, including the following [3, 4]:

a. **Roulette Wheel Selection** (also known as **Fitness Proportionate Selection**) – each individual of the population is assigned a portion of a roulette wheel, where a larger portion is allocated to the individual with a higher fitness value. Then, the roulette wheel is continuously rotated until the required amount of parent chromosomes has been selected.

b. **Stochastic Universal Sampling** – each individual of the population is assigned a portion of a straight line segment, where a larger portion is allocated to the individual with a higher fitness value (similar to the Roulette Wheel Selection mechanism). Then, the parent chromosomes are selected based on the evenly spaced fitness intervals (unlike Roulette Wheel Selection, which requires generating a random number each time in order to rotate the roulette wheel).

c. **Binary Tournament Selection** – multiple binary tournaments are executed, where two individuals are randomly sampled from the population during each tournament, and the individual with a higher fitness value is chosen to become a parent. The required number of tournaments is determined based on the population size.

d. **Ranking Selection** – the parent and offspring chromosomes from the previous generation are combined in a one data structure, sorted based on their fitness, and a subset of chromosomes with higher fitness values (out of the available parent and offspring chromosomes) will become parents. Such selection mechanism has been widely used in canonical Evolutionary Strategies [3] and is generally referred to as $(\mu + \lambda)$-selection, where parents (μ) are allowed to compete with offspring (λ).

e.   **Uniform Sampling** – the parent chromosomes are selected from the population by uniform (or random) sampling. Unlike the aforementioned selection mechanisms, Uniform Sampling is not biased by fitness.

For a detailed description of the considered non-parametric selection mechanisms and illustrative examples of these mechanisms, this study refers to Eiben and Smith [3] and Sivanandam and Deepa [4]. Five categories of the EA algorithm, deploying different types of parent selection mechanisms, will be evaluated in this study, including the following: (1) EA with Roulette Wheel Selection (EA-RWS); (2) EA with Stochastic Universal Sampling (EA-SUS); (3) EA with Binary Tournament Selection (EA-BTS); (4) EA with Ranking Selection (EA-RS); and (5) EA with Uniform Sampling (EA-US).

4.6. EA operations

Once the parent chromosomes are selected, the developed EA algorithm applies the crossover and mutation operators in order to produce and mutate the offspring chromosomes. Both operators are described in sections 4.6.1–4.6.2 of the chapter.

4.6.1. Crossover

The order crossover is used to produce the offspring chromosomes. Selection of the latter crossover operator can be justified by the adopted chromosome representation. Specifically, certain crossover operators (e.g., N-point, whole arithmetic, uniform) may produce infeasible offspring for the integer chromosome representation [3]. On the other hand, the order crossover guarantees feasibility of the generated offspring chromosomes. An example of an order crossover operation is illustrated in **Figure 3**. Two chromosomes are randomly selected from the available parent chromosomes. The probability of parents to undergo a crossover operation is determined by the crossover probability parameter (σ^C). After that, a string of genes is copied from parent "1" to offspring "1". Note that the length of a string will be set randomly, and, therefore, may vary from one crossover operation to another. In the considered example, a string of genes with jobs "2", "6", "8", and "3" is copied from parent "1" to offspring "1". Then, the genes with missing jobs are copied from parent "2" to offspring "1". In the

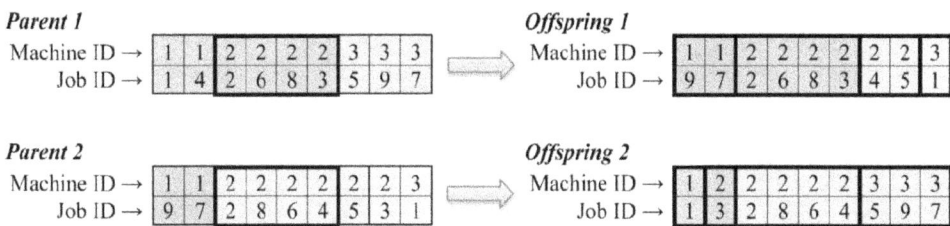

Figure 3. Order crossover operation example.

Before Swap Mutation

Machine ID →	1	1	1	2	2	2	3	3	3
Job ID →	2	3	5	4	6	9	1	7	8

Before Insert Mutation

Machine ID →	1	1	1	2	2	2	3	3	3
Job ID →	2	3	5	4	6	9	1	7	8

After Swap Mutation

Machine ID →	1	1	1	2	2	2	3	3	3
Job ID →	7	3	5	4	6	9	1	2	8

After Insert Mutation

Machine ID →	1	1	1	1	2	2	2	3	3
Job ID →	2	4	3	5	6	1	9	7	8

Figure 4. Mutation operation example.

considered example, jobs "9", "7", "4", "5", and "1" are copied from parent "2" to offspring "1". The offspring "2" is produced in a similar manner.

4.6.2. Mutation

The offspring chromosomes, produced via the order crossover, will be mutated. Two types of mutation operators will be applied in this study: (a) swap; and (b) insert. An example of a mutation operation is illustrated in **Figure 4**. In case of a swap mutation operation, job "2", initially scheduled for processing on machine "1" as the first job, is re-scheduled for processing on machine "3" as the second job. On the other hand, job "7", initially scheduled for processing on machine "3" as the second job, is re-scheduled for processing on machine "1" as the first job. In case of an insert mutation operation, job "4", initially scheduled for processing on machine "2" as the first job, is re-scheduled for processing on machine "1" as the second job. On the other hand, job "1", initially scheduled for processing on machine "3" as the first job, is re-scheduled for processing on machine "2" as the second job. Application of both swap and insert mutation operators allows altering job to machine and job to processing order assignments. The number of genes to be mutated throughout the mutation operation is determined by the mutation probability parameter (σ^M).

4.7. Offspring selection mechanism

The purpose of the offspring selection mechanism is to determine a subset of individuals from the generated offspring chromosomes that will survive in the given generation and will be moved to the next generation. This study relies on the generational offspring selection mechanism, where all offspring chromosomes will be moved to the next generation and become candidate parent chromosomes. Such offspring selection mechanism has been widely used in canonical Genetic Algorithms, proposed by Holland, and Genetic Programming, developed by Koza [3, 4].

4.8. Stopping criterion

The developed EA algorithm will be terminated, once a certain stopping criterion is met. The stopping criterion, adopted in this study, is the maximum number of generations (g^{MAX}).

5. Computational experiments

This section provides a detailed description of the computational experiments, which were conducted to evaluate the considered non-parametric selection mechanisms. Five EA categories, applying different non-parametric selection mechanisms (i.e., the EA-RWS, EA-SUS, EA-BTS, EA-RS, and EA-US algorithms, described in Section 4.5), were evaluated in terms of the objective function value at termination, computational time, and changes in the population diversity throughout the algorithmic run. All EA algorithms were coded in MATLAB 2016a. The computational experiments were executed on a CPU with Dell Intel(R) Core™ i7 Processor and 32 GB of RAM. Sections 5.1–5.3 elaborate on the input data selection for the MSP mathematical model, parameter tuning of the developed EA algorithms, and comprehensive comparative analysis of the considered non-parametric selection mechanisms.

5.1. Input data selection

The required input parameters for the MSP mathematical model were primarily generated based on the relevant literature [2, 6–36]. The adopted parameter values are presented in **Table 1**. A total of 40 problem instances were developed to conduct the computational experiments by changing the number of arriving jobs from 50 to 140 with an increment of 10 jobs, while the number of available machines was changed from 4 to 10 with an increment of 2 machines.

MSP parameter	Adopted value
Number of arriving jobs: m (jobs)	Varies based on the problem instance
Number of available machines: n (machines)	Varies based on the problem instance
Number of job processing orders: p (orders)	$p = m$ (considering the case, when all jobs are assigned for processing on one machine)
Arrival time of job i: $AT_i, i \in I$ (hours)	$Exponential(2)/60$
Handling time of job i on machine j: $HT_{ij}, i \in I, j \in J$ (hours)	$Uniform(20;80)/60$
Deadline for processing job i: $DP_i, i \in I$ (hours)	$AT_i + Uniform(1.2;1.5) \cdot \min_{j \in J}(HT_{ij})$
Unit handling cost for job i: $c_i^{HC}, i \in I$ (USD/hour)	$Uniform(200;400)$
Unit waiting cost for job i: $c_i^{WC}, i \in I$ (USD/hour)	$Uniform(50;100)$
Unit delayed processing cost of job i: $c_i^{DC}, i \in I$ (USD/hour)	$Uniform(300;600)$
Large positive number: PN	10^6

Exponential(a) – exponentially distributed pseudorandom numbers with a mean inter-arrival time of a; *Uniform(b;c)* – uniformly distributed pseudorandom numbers, varying from b to c.

Table 1. MSP parameter values.

5.2. EA parameter tuning

A parameter selection analysis was performed for the EA-RWS, EA-SUS, EA-BTS, EA-RS, and EA-US algorithms to identify the appropriate parameter values. Each one of the developed EA algorithms has a total of 4 parameters, including the following: (1) population size – Ω; (2) crossover probability – σ^C; (3) mutation probability – σ^M; and (4) maximum number of generations – g^{MAX}. A "full factorial design" methodology [5], described in the introduction section of the chapter, was adopted for the EA parameter tuning. A total of 3 candidate values were considered for each parameter (i.e., 3^f factorial design). A total of 3 problem instances were chosen at random from the generated problem instances (see Section 5.1) in order to conduct the parameter tuning analysis.

A total of 10 replications were performed for each algorithm and each problem instance to obtain the average objective function and computational time values. The number of replications was found to be sufficient, as the objective function values did not vary substantially from one replication to another. Specifically, the coefficient of variation of the objective function values at termination did not exceed 1.00% over the performed replications for all the generated problem instances and the developed solution algorithms. Based on preliminary algorithmic runs, it was found that increasing number of replications would incur a significant increase in the computational time without a significant reduction of the objective function coefficient of variation for each EA. **Table 2** provides a summary of the parameter analysis for each EA, including the following data: (1) algorithm; 2) parameter; (3) considered candidate values for each parameter; and (4) the best parameter value, highlighted in bold font (determined based on the analysis of a tradeoff between the obtained objective function values and computational time required).

The parameter tuning analysis for the developed EA algorithms took more than 11 days (i.e., more than 51 hours for each EA). Application of parametric selection mechanisms would increase the computational time of the parameter tuning analysis even further. The latter highlights the importance of adopting non-parametric selection mechanisms.

5.3. Comparative analysis

This section focuses on a detailed comparative analysis of the considered EA algorithms, deploying different non-parametric selection mechanisms, in terms of the objective function

Algorithm\Parameter	Ω	σ^C	σ^M	g^{MAX}
EA-RWS	[40; 50; **60**]	[**0.25**; 0.50; 0.75]	[0.01; 0.02; 0.05]	[2000; 2500; **3000**]
EA-SUS	[40; 50; **60**]	[0.25; 0.50; **0.75**]	[0.01; 0.02; 0.05]	[2000; 2500; **3000**]
EA-BTS	[40; 50; **60**]	[0.25; 0.50; **0.75**]	[**0.01**; 0.02; 0.05]	[2000; 2500; **3000**]
EA-RS	[40; 50; **60**]	[0.25; 0.50; **0.75**]	[0.01; **0.02**; 0.05]	[2000; 2500; **3000**]
EA-US	[40; 50; **60**]	[0.25; 0.50; **0.75**]	[**0.01**; 0.02; 0.05]	[2000; 2500; **3000**]

Table 2. EA parameter tuning analysis summary.

values at termination and required computational time. Moreover, changes in the population diversity are analyzed throughout evolution of each EA.

5.3.1. Objective function and computational time

The developed EA-RWS, EA-SUS, EA-BTS, EA-RS, and EA-US algorithms were executed for all the generated problem instances, which were described in Section 5.1. A total of 10 replications were performed for each algorithm and each problem instance. Results of the conducted analysis are reported in **Table 3** for each algorithm and each problem instance, including the following data: (1) instance number; (2) number of arriving jobs (m); (3) number of available machines (n); (4) average objective function value at termination (Z) for each EA algorithm; and (5) average computational time value (CPU) for each EA algorithm.

The average objective function values comprised 339.79 10^3 USD, 321.39 10^3 USD, 333.97 10^3 USD, and 324.14 10^3 USD, and 357.86 10^3 USD over the developed problem instances for the EA-RWS, EA-SUS, EA-BTS, EA-RS, and EA-US algorithms respectively. Therefore, EA-SUS that relies on Stochastic Universal Sampling outperformed the EAs with other non-parametric selection mechanisms in terms of the solution quality. Superiority of the EA-SUS algorithm can be explained by the fact that Stochastic Universal Sampling selects the parent chromosomes based on the evenly distributed fitness intervals and, therefore, ensures that high, medium, and low quality individuals will be given a chance to reproduce. The EA-RS algorithm, which deploys Ranking Selection, demonstrated a good performance; however, it was outperformed by the EA-SUS algorithm due to the fact that ranking is substantially biased by fitness. Ranking Selection allows only high and medium fitness chromosomes to become parents, while the individuals with low fitness values are not given any chance to reproduce.

The EA-RWS and EA-BTS algorithms were outperformed by both EA-SUS and EA-RS algorithms, as they do not guarantee that high and medium quality individuals will become parents. Although Roulette Wheel Selection and Binary Tournament Selection are biased by fitness, and the individuals with higher fitness have higher chances to reproduce, such selection mechanisms may allow a significant portion of low quality individuals to become parents, which negatively affects the objective function values and results in a premature convergence. The worst performance was recorded for the EA-US algorithm, which relies on Uniform Sampling. Uniform Sampling is not biased by fitness and gives all individuals equal chances to become parents, which may not be desirable in some cases (i.e., higher and medium quality individuals should have higher chances to reproduce, as compared to low quality individuals). Uniform Sampling can be advantageous when applied in combination with other selection mechanisms (e.g., Uniform Sampling is used at the parent selection stage, while Stochastic Universal Sampling is used at the offspring selection stage). Evaluation of the EA algorithms, which use a combination of various non-parametric selection mechanisms, will be one of the future research directions of this study.

An additional statistical analysis was conducted to investigate differences between the average objective function values at termination, suggested by the developed algorithms. The null hypothesis was assumed to be $H_0 : \mu_{EA_1} = \mu_{EA_2}$ (i.e., the average objective function value at

Instance	m	n	EA-RWS		EA-SUS		EA-BTS		EA-RS		EA-US	
			Z, 10^3 USD	CPU, sec	Z, 10^3 USD	CPU, sec	Z, 10^3 USD	CPU, sec	Z, 10^3 USD	CPU, sec	Z, 10^3 USD	CPU, sec
1	50	4	141.61	52.72	137.06	56.26	138.30	56.61	138.86	55.11	143.75	61.19
2	50	6	84.43	51.76	82.89	53.53	82.94	58.46	87.32	54.45	89.29	62.14
3	50	8	59.35	52.92	56.85	54.81	57.81	59.75	57.84	55.42	61.45	63.01
4	50	10	44.52	54.54	42.98	56.07	45.19	60.40	45.14	56.32	48.43	63.94
5	60	4	198.97	58.39	190.43	60.07	191.37	65.79	192.01	60.47	195.31	68.20
6	60	6	121.49	59.69	111.83	61.33	114.81	67.33	113.47	62.06	123.11	69.39
7	60	8	82.89	61.17	80.00	62.16	80.71	68.59	80.60	63.14	84.01	70.83
8	60	10	62.31	62.12	60.37	63.35	62.98	70.17	60.61	64.00	66.13	72.39
9	70	4	278.59	65.71	259.23	67.40	269.27	74.93	267.85	68.11	285.21	77.10
10	70	6	164.86	67.19	159.63	69.17	161.25	76.34	159.87	69.38	176.77	78.77
11	70	8	119.25	68.20	111.44	70.07	114.07	77.85	113.52	71.61	122.13	80.20
12	70	10	87.04	69.71	84.72	70.99	88.47	79.31	85.13	72.72	92.40	82.75
13	80	4	358.82	73.98	341.87	75.49	347.55	84.01	342.51	77.36	368.74	87.29
14	80	6	214.52	75.25	204.72	76.47	212.32	84.90	206.70	78.60	222.90	88.00
15	80	8	148.00	76.39	142.60	79.78	148.80	85.86	143.65	79.50	155.65	89.31
16	80	10	112.84	77.59	106.88	79.47	110.21	87.52	107.23	80.47	124.04	91.83
17	90	4	460.90	81.58	444.03	83.76	446.23	92.98	446.81	85.02	484.48	96.91
18	90	6	277.98	82.84	269.76	84.89	271.57	93.84	271.19	86.96	297.47	92.80
19	90	8	191.22	83.95	180.81	86.22	195.22	95.37	180.97	88.43	203.99	94.31
20	90	10	151.94	85.54	135.86	87.94	142.50	97.32	136.17	89.75	159.42	95.93
21	100	4	600.06	89.07	564.90	92.29	580.63	101.97	568.24	94.03	601.43	101.61
22	100	6	355.99	89.84	343.57	93.10	348.84	101.66	346.33	94.63	384.70	102.99
23	100	8	249.85	91.11	228.49	94.42	243.06	103.32	229.37	95.94	260.30	103.12
24	100	10	190.11	92.81	171.88	95.95	184.94	104.81	174.24	97.51	196.91	104.42
25	110	4	720.16	96.39	678.10	99.21	706.05	109.70	678.90	101.52	745.49	108.94
26	110	6	440.85	98.03	419.34	101.31	429.49	111.74	421.62	102.51	461.96	110.18
27	110	8	300.11	99.59	280.76	102.57	292.38	112.76	281.59	104.07	317.98	111.86
28	110	10	223.92	100.81	208.87	103.90	220.49	114.61	210.76	105.66	245.74	113.24
29	120	4	858.23	104.78	802.81	108.03	848.19	120.34	816.61	110.30	900.09	120.61
30	120	6	539.24	105.52	488.90	109.46	514.46	120.73	498.86	112.01	549.63	122.26
31	120	8	356.24	107.50	343.26	111.21	363.44	122.90	345.24	113.77	389.83	123.98
32	120	10	273.65	109.06	249.24	112.04	267.37	125.11	250.87	114.76	284.92	124.56
33	130	4	1011.56	112.84	974.07	116.21	1001.36	123.02	979.15	119.14	1069.10	129.10
34	130	6	620.42	114.00	583.82	117.26	618.10	122.51	589.85	119.96	650.25	130.27

Instance	m	n	EA-RWS		EA-SUS		EA-BTS		EA-RS		EA-US	
			$Z, 10^3$ USD	CPU, sec	$Z, 10^3$ USD	CPU, sec	$Z, 10^3$ USD	CPU, sec	$Z, 10^3$ USD	CPU, sec	$Z, 10^3$ USD	CPU, sec
35	130	8	418.78	115.49	401.15	118.47	424.44	124.11	403.36	121.57	449.19	132.76
36	130	10	310.32	116.52	296.11	119.97	318.84	125.34	298.81	122.95	342.26	134.08
37	140	4	1184.91	120.74	1121.27	123.96	1168.76	129.81	1123.67	127.12	1267.22	138.33
38	140	6	712.51	121.49	680.00	125.12	696.63	130.66	690.01	129.63	757.52	139.52
39	140	8	499.65	123.61	465.71	126.37	483.23	131.65	469.30	129.50	529.86	141.40
40	140	10	363.36	124.70	349.41	127.91	366.61	133.38	351.45	130.79	405.48	142.81
Average:			339.79	87.38	321.39	89.95	333.97	97.69	324.14	91.66	357.86	100.56

Table 3. Objective function and computational time values for the considered EA algorithms.

termination of algorithm EA_1 [μ_{EA_1}] is equal to the average objective function value at termination of algorithm EA_2 [μ_{EA_2}]), while the alternative hypothesis was assumed to be $H_a : \mu_{EA_1} < \mu_{EA_2}$ (algorithm EA_1 returns lower average objective function value at termination as compared to algorithm EA_2). The average objective function values were estimated over 40 problem instances for each EA algorithm. Based on the hypothesis testing results, no statistically significant difference has been identified among the average objective function values at termination, suggested by the EA-SUS algorithm and other developed EA algorithms, at significance level $\alpha = 0.05$. The latter finding can be justified by the fact that for some of the problem instances the developed algorithms did not demonstrate significant differences in terms of the objective function values (generally, the problem instances with lower number of arriving jobs and available machines – problem instances 1, 2, 5, 6, and others).

Furthermore, on average over all the generated problem instances the EA-SUS algorithm outperformed the EA-RWS, EA-BTS, EA-RS, and EA-US algorithms by 5.72, 3.91, 0.86, and 11.35%. However, for some of the problem instances the EA-SUS algorithm outperformed the EA-RWS, EA-BTS, EA-RS, and EA-US algorithms by up to 11.84, 7.97, 5.34, and 17.65%. Therefore, application of the EA-SUS algorithm is expected to become even more advantageous (in terms of objective function values at termination) with increasing problem size. The computational time of the developed EA algorithms did not exceed 142.81 sec over all 40 problem instances, which can be considered as acceptable.

5.3.2. Changes in the population diversity

The population diversity is critical in EAs especially at early stages of the search process. Without a diverse population, a given EA will not be able explore the available domains of the search space in an efficient manner. Lack of diversity in early generations of the EA algorithm may lead to negative consequences, including premature convergence. The population fitness values were recorded throughout evolution of the developed EA-RWS, EA-SUS,

EA-BTS, EA-RS, and EA-US algorithms for each replication and each problem instance. The population fitness boxplots are illustrated in **Figures 5** and **6** for the first replication of each EA algorithm after the parent selection in generations 500, 1000, 1500, 2000, 2500, and 3000. Note that boxplots are presented only for the first replication of each EA algorithm and problem instances 37–40 (i.e., the problem instances with the largest number of arriving jobs), but

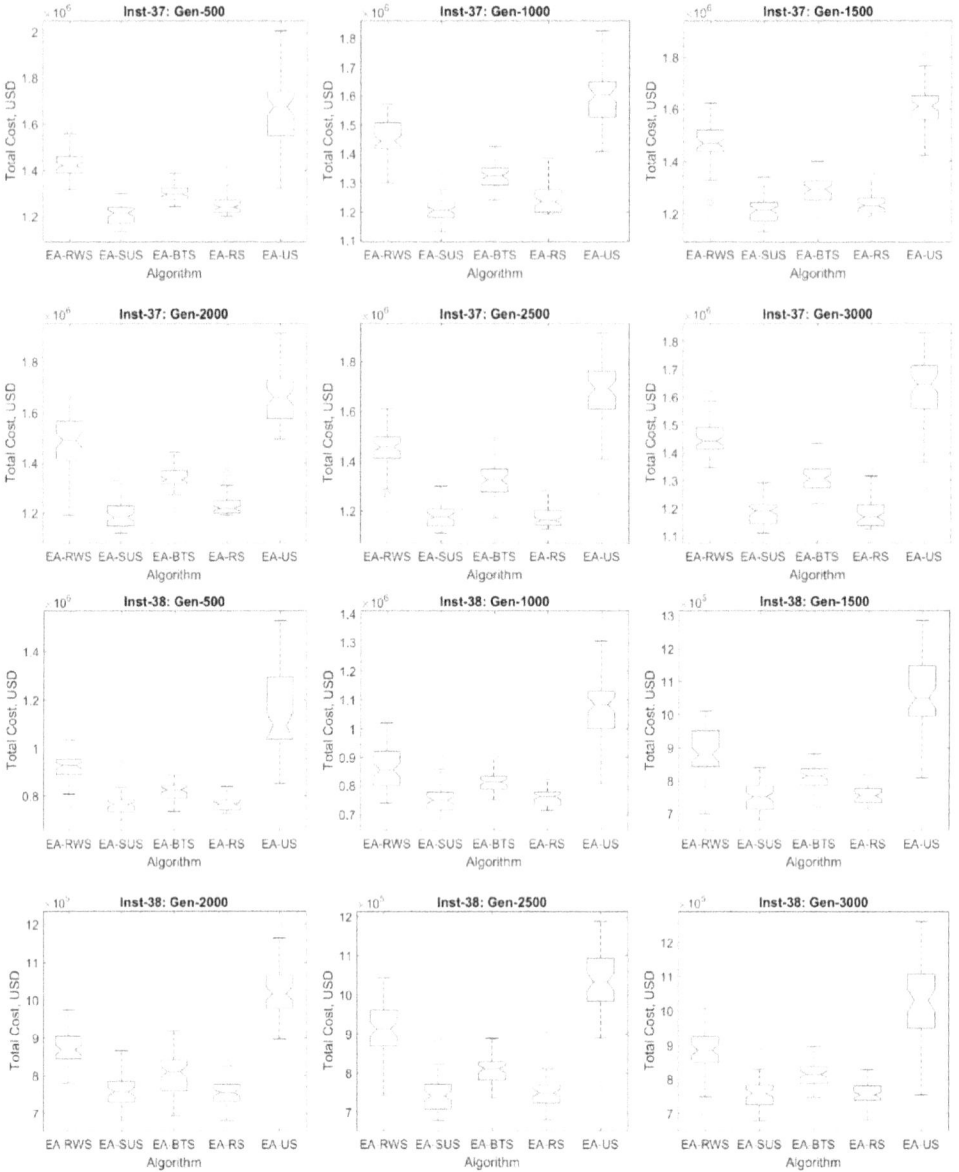

Figure 5. EA population fitness boxplots for problem instances 37 and 38.

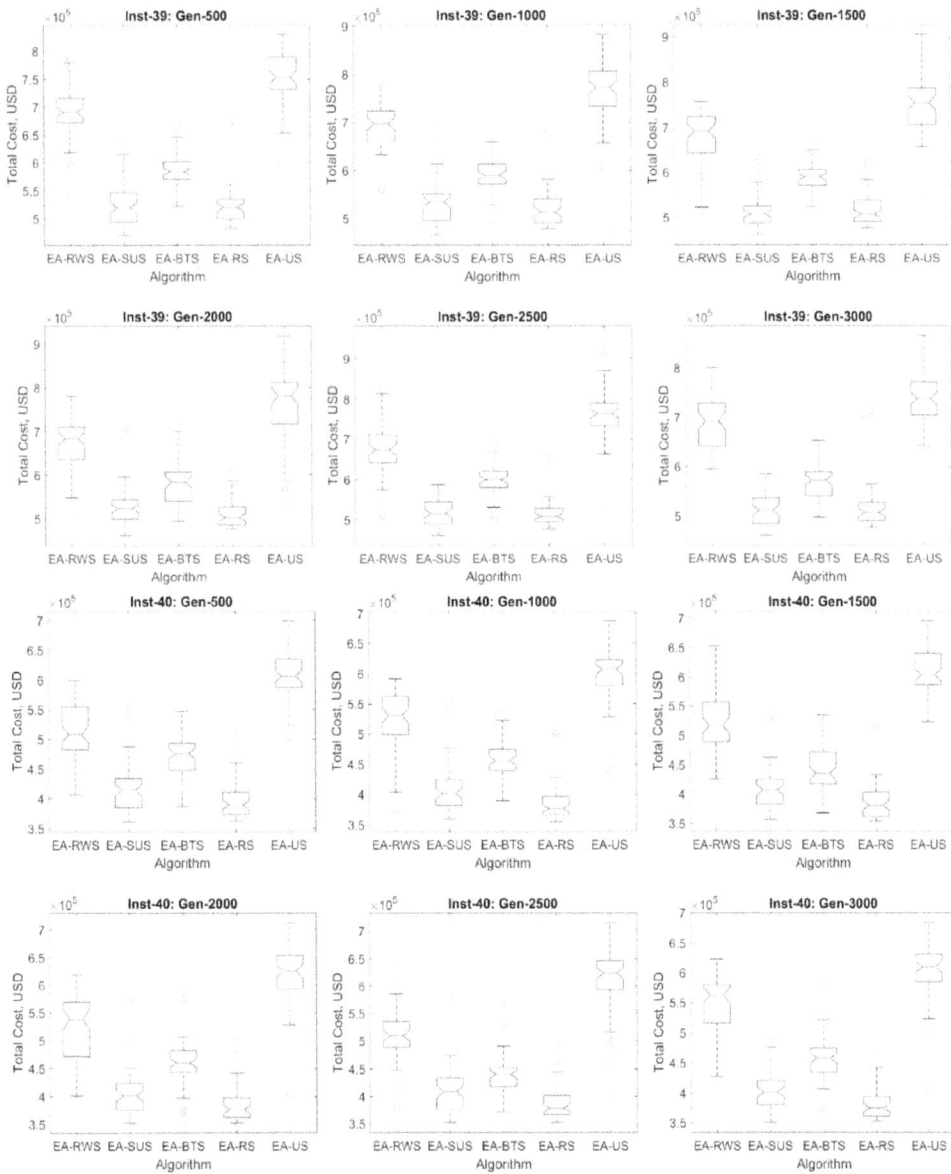

Figure 6. EA population fitness boxplots for problem instances 39 and 40.

similar patterns have been observed for the rest of algorithmic replications and problem instances. The population fitness boxplots have the following components: (1) rectangle, where the top and the bottom parts correspond to 75th and 25th population fitness value percentiles respectively; (2) median, which is shown using a red line; (3) whiskers, which are shown using dashed lines covering 99.30% of the population fitness value data points; and (4) extreme population fitness value points (falling outside of 99.30% of the population fitness value data points) or "outliers", which are shown using "°" symbol.

It can be observed that the population fitness boxplot whiskers of the EA-RWS and EA-US algorithms consistently cover a wider range of the population fitness values. The latter finding indicates that both EA-RWS and EA-US algorithms maintain a more diverse population, as compared to the EA-SUS, EA-BTS, and EA-RS algorithms. However, the quality of individuals within both EA-RWS and EA-US populations is significantly lower as compared to the EA-SUS, EA-BTS, and EA-RS populations. For example, the EA-RWS and EA-US algorithms cover the population fitness ranges of $[1193.15; 1627.94]$ 10^3 USD and $[1276.65; 1829.78]$ 10^3 USD respectively, while the EA-SUS algorithm covers the population fitness range of $[1110.66; 1387.16]$ 10^3 USD for problem instance 37 at termination (i.e., in generation 3000). Therefore, as discussed in Section 5.3.1, the EA-RWS and EA-US algorithms were outperformed by the EA-SUS, EA-BTS, and EA-RS algorithms in terms of the objective function values at termination. The EA-SUS, EA-BTS, and EA-RS algorithms were able to maintain the adequate population diversity and return good quality job to machine to processing order assignments.

Throughout the computational experiments, it was found that the population diversity patterns did not change significantly from generation 500 up to generation 3000 (e.g., the range, covered by the population fitness boxplot whiskers, does not alter substantially throughout evolution of each EA after generation 500). The latter finding can be justified by the fact that the developed EAs relatively quickly identified the promising domains of the search space (i.e., within the first 400–500 generations), and then focused on exploiting the identified domains for the rest of generations, aiming to discover solutions with superior fitness values. Application of scaling mechanisms (such as linear scaling, sigma truncation, and power law scaling) will allow controlling the population diversity of the developed EA algorithms (e.g., reduce the population diversity towards the EA convergence and give higher reproduction chances to "super-individuals" – i.e. the individuals with the highest fitness values) and will be one of the future research directions of this study.

6. Concluding remarks and future research extensions

Evolutionary Algorithms and other metaheuristic algorithms have been extensively applied for solving complex stochastic, robust, and dynamic optimization problems. Two types of selection mechanisms are deployed within Evolutionary Algorithms, including the parent selection and the offspring selection. Evolutionary Algorithms have a lot of parameters, which are generally set based on the parameter tuning analysis. Parametric selection mechanisms (e.g., Exponential Ranking Selection, Tournament Selection, Boltzmann Selection) increase the number of parameters within a given Evolutionary Algorithm, which can make the parameter tuning analysis computationally prohibitive due to significant computational time required. To avoid the latter drawback and facilitate the parameter tuning analysis of Evolutionary Algorithms, this study focused on design of the Evolutionary Algorithm that solely relied on non-parametric selection mechanisms. Different categories of Evolutionary Algorithms, which applied various non-parametric selection mechanisms (Roulette Wheel Selection, Stochastic Universal Sampling, Binary Tournament Selection, Ranking Selection, Uniform Sampling), were evaluated based on the major algorithmic performance indicators.

A set of computational experiments were conducted for the unrelated machine scheduling problem, which is known to be NP-hard. The objective of the mathematical model, proposed for the problem, aimed to minimize the total job processing cost. Results indicate that the Evolutionary Algorithm with the Stochastic Universal Sampling selection mechanism outperforms the Evolutionary Algorithms with other selection mechanisms in terms of the objective function values. The worst performance was demonstrated by the Evolutionary Algorithm, which relied on the Uniform Sampling selection mechanism. Furthermore, the Evolutionary Algorithms with the Roulette Wheel Selection and Uniform Sampling selection mechanisms typically allowed maintaining higher population diversity; however, the quality of individuals within the population was lower as compared to the Evolutionary Algorithms with the Stochastic Universal Sampling, Binary Tournament Selection, and Ranking Selection mechanisms. The computational time of all the developed Evolutionary Algorithms did not exceed 142.81 sec over the considered problem instances, which can be considered as acceptable. Therefore, based on a comprehensive analysis of the commonly used non-parametric selection mechanisms, Stochastic Universal Sampling was found to be the most promising, as it was able to maintain the adequate population diversity throughout the algorithmic run and return good quality solutions at termination. Results from the conducted numerical experiments are expected to facilitate development of efficient Evolutionary Algorithms for the machine scheduling problems. Moreover, the developed problem instances and findings from this study can serve as benchmarks for the future machine scheduling studies.

The future research directions for this study include the following: (1) application of scaling mechanisms for the fitness function; (2) evaluation of the Evolutionary Algorithms, which use a combination of various non-parametric selection mechanisms (e.g., Uniform Sampling is used at the parent selection stage, while Stochastic Universal Sampling is used at the offspring selection stage); (3) consider alternative stopping criteria for the developed Evolutionary Algorithms; (4) compare various non-parametric selection mechanisms for the Hybrid Evolutionary Algorithms, which apply different local search heuristics along with the stochastic search operators; and (5) evaluate performance of the commonly used non-parametric selection mechanisms for other NP-hard problems (e.g., bin packing problem, Knapsack problem, traveling salesman problem).

Nomenclature

Sets

$I = \{1, ..., m\}$ set of arriving jobs

$J = \{1, ..., n\}$ set of available machines

$K = \{1, ..., p\}$ set of job processing orders

Decision variables

$x_{ijk} \in \{0, 1\} \forall i \in I, j \in J, k \in K$ =1 if arriving job i is scheduled for processing on machine j in processing order k (=0 otherwise)

Auxiliary variables

$IT_{ijk} \in R^+ \forall i \in I, j \in J, \; k \in K$ idling time of machine j between processing job i and preceding job processed in order $(k-1)$ (hours)

$SPT_i \in R^+ \forall i \in I$ start processing time for job i (hours)

$FPT_i \in R^+ \forall i \in I$ finish processing time for job i (hours)

$WT_i \in R^+ \forall i \in I$ waiting time of job i (hours)

$PD_i \in R^+ \forall i \in I$ delay in processing job i (hours)

Parameters

$m \in N$ number of arriving jobs (jobs)

$n \in N$ number of available machines (machines)

$p \in N$ number of job processing orders (orders)

$AT_i \in R^+ \forall i \in I$ arrival time of job i (hours)

$HT_{ij} \in R^+ \forall i \in I, j \in J$ handling time of job i on machine j (hours)

$DP_i \in R^+ \forall i \in I$ deadline for processing job i (hours)

$c_i^{HC} \in R^+ \forall i \in I$ unit handling cost for job i (USD/hour)

$c_i^{WC} \in R^+ \forall i \in I$ unit waiting cost for job i (USD/hour)

$c_i^{DC} \in R^+ \forall i \in I$ unit delayed processing cost of job i (USD/hour)

$PN \in R^+$ large positive number

Author details

Maxim A. Dulebenets

Address all correspondence to: mdulebenets@eng.famu.fsu.edu

Department of Civil and Environmental Engineering, Florida A&M University-Florida State University, Tallahassee, FL, USA

References

[1] Hromkovič J. Algorithmics for Hard Problems: Introduction to Combinatorial Optimization, Randomization, Approximation, and Heuristics. 2nd ed. Berlin, Germany: Springer International Publishing; 2002. p. 557. DOI: 10.1007/978-3-662-05269-3

[2] Pinedo M. Scheduling: Theory, Algorithms, and Systems. 5th ed. New York, USA: Springer International Publishing; 2016. p. 670. DOI: 10.1007/978-3-319-26580-3

[3] Eiben AE, Smith JE. Introduction to Evolutionary Computing. 2nd ed. Berlin, Germany: Springer International Publishing; 2015. p. 287. DOI: 10.1007/978-3-662-44874-8

[4] Sivanandam SN, Deepa SN. Introduction to Genetic Algorithms. 1st ed. Berlin, Germany: Springer International Publishing; 2008. p. 442. DOI: 10.1007/978-3-540-73190-0

[5] de Lima EB, Pappa GL, de Almeida JM, Gonçalves MA, Meira W. Tuning Genetic Programming parameters with factorial designs. In: Proceedings of the IEEE Congress on Evolutionary Computation (CEC); 18–23 July 2010; Barcelona, Spain. New York: IEEE; 2010. pp. 1-8

[6] Boysen N, Briskorn D, Meisel F. A generalized classification scheme for crane scheduling with interference. European Journal of Operational Research. 2017;258(1):343-357. DOI: 10.1016/j.ejor.2016.08.041

[7] Nagananda KG, Khargonekar P. An approximately optimal algorithm for scheduling phasor data transmissions in smart grid networks. IEEE Transactions on Smart Grid. 2017;8(4):1649-1657. DOI: 10.1109/TSG.2015.2497284

[8] Fernandez-Viagas V, Ruiz R, Framinan JM. A new vision of approximate methods for the permutation flowshop to minimise makespan: State-of-the-art and computational evaluation. European Journal of Operational Research. 2017;257(3):707-721. DOI: 10.1016/j.ejor.2016.09.055

[9] Ozturk O, Chu C. Exact and metaheuristic algorithms to minimize the total tardiness of cutting tool sharpening operations. Expert Systems with Applications. 2018;95:224-235. DOI: 10.1016/j.eswa.2017.11.030

[10] Juarez F, Ejarque J, Badia RM. Dynamic energy-aware scheduling for parallel task-based application in cloud computing. Future Generation Computer Systems. 2018;78:257-271. DOI: 10.1016/j.future.2016.06.029

[11] Dulebenets MA. Application of evolutionary computation for berth scheduling at marine container terminals: Parameter tuning versus parameter control. IEEE Transactions on Intelligent Transportation Systems. 2018;19(1):25-37. DOI: 10.1109/TITS.2017.2688132

[12] Herrmann J, Proth JM, Sauer N. Heuristics for unrelated machine scheduling with precedence constraints. European Journal of Operational Research. 1997;102(3):528-537. DOI: 10.1016/S0377-2217(96)00247-0

[13] Weng MX, Lu J, Ren H. Unrelated parallel machine scheduling with setup consideration and a total weighted completion time objective. International Journal of Production Economics. 2001;**70**(3):215-226. DOI: 10.1016/S0925-5273(00)00066-9

[14] Vallada E, Ruiz R. A genetic algorithm for the unrelated parallel machine scheduling problem with sequence dependent setup times. European Journal of Operational Research. 2011;**211**(3):612-622. DOI: 10.1016/j.ejor.2011.01.011

[15] Bank J, Werner F. Heuristic algorithms for unrelated parallel machine scheduling with a common due date, release dates, and linear earliness and tardiness penalties. Mathematical and Computer Modelling. 2001;**33**(4):363-383. DOI: 10.1016/S0895-7177(00)00250-8

[16] Glass CA, Potts CN, Shade P. Unrelated parallel machine scheduling using local search. Mathematical and Computer Modelling. 1994;**20**(2):41-52. DOI: 10.1016/0895-7177(94)90205-4

[17] Pearn WL, Chung SH, Yang MH, Chen YH. Algorithms for the wafer probing scheduling problem with sequence-dependent set-up time and due date restrictions. Journal of the Operational Research Society. 2004;**55**(11):1194-1207. DOI: 10.1057/palgrave.jors.2601795

[18] Rabadi G, Moraga RJ, Al-Salem A. Heuristics for the unrelated parallel machine scheduling problem with setup times. Journal of Intelligent Manufacturing. 2006;**17**(1):85-97. DOI: 10.1007%2Fs10845-005-5514-0

[19] Kim DW, Na DG, Chen FF. Unrelated parallel machine scheduling with setup times and a total weighted tardiness objective. Robotics and Computer-Integrated Manufacturing. 2003;**19**(1):173-181. DOI: 10.1016/S0736-5845(02)00077-7

[20] Aspnes J, Azar Y, Fiat A, Plotkin S, Waarts O. On-line routing of virtual circuits with applications to load balancing and machine scheduling. Journal of the ACM (JACM). 1997;**44**(3):486-504. DOI: 10.1145/258128.258201

[21] Hsieh JC, Chang PC, Hsu LC. Scheduling of drilling operations in printed circuit board factory. Computers and Industrial Engineering. 2003;**44**(3):461-473. DOI: 10.1016/S0360-8352(02)00231-0

[22] Chen JF, Wu TH. Total tardiness minimization on unrelated parallel machine scheduling with auxiliary equipment constraints. Omega. 2006;**34**(1):81-89. DOI: 10.1016/j.omega.2004.07.023

[23] Jinsong B, Xiaofeng H, Ye J. A genetic algorithm for minimizing makespan of block erection in shipbuilding. Journal of Manufacturing Technology Management. 2009;**20**(4):500-512. DOI: 10.1108/17410380910953757

[24] Agnetis A, Flamini M, Nicosia G, Pacifici A. Scheduling three chains on two parallel machines. European Journal of Operational Research. 2010;**202**(3):669-674. DOI: 10.1016/j.ejor.2009.07.001

[25] Hu X, Bao JS, Jin Y. Minimising makespan on parallel machines with precedence constraints and machine eligibility restrictions. International Journal of Production Research. 2010;48(6):1639-1651. DOI: 10.1080/00207540802620779

[26] Driessel R, Mönch L. Variable neighborhood search approaches for scheduling jobs on parallel machines with sequence-dependent setup times, precedence constraints, and ready times. Computers and Industrial Engineering. 2011;61(2):336-345. DOI: 10.1016/j.cie.2010.07.001

[27] Agnetis A, Kellerer H, Nicosia G, Pacifici A. Parallel dedicated machines scheduling with chain precedence constraints. European Journal of Operational Research. 2012;221(2):296-305. DOI: 10.1016/j.ejor.2012.03.040

[28] Park C, Seo J. A GRASP approach to transporter scheduling and routing at a shipyard. Computers & Industrial Engineering. 2012;63(2):390-399. DOI: 10.1016/j.cie.2012.04.010

[29] Park C, Seo J. A GRASP approach to transporter scheduling for ship assembly block operations management. European Journal of Industrial Engineering. 2013;7(3):312-332. DOI: 10.1504/EJIE.2013.054133

[30] Rose CD, Coenen JM. Comparing four metaheuristics for solving a constraint satisfaction problem for ship outfitting scheduling. International Journal of Production Research. 2015;53(19):5782-5796. DOI: 10.1080/00207543.2014.998786

[31] Nicosia G, Pacifici A. Scheduling assembly tasks with caterpillar precedence constraints on dedicated machines. International Journal of Production Research. 2017;55(6):1680-1691. DOI: 10.1080/00207543.2016.1220686

[32] Dulebenets MA. The vessel scheduling problem in a liner shipping route with heterogeneous vessel fleet. International Journal of Civil Engineering. 2018;16(1):19-32. DOI: 10.1007/s40999-016-0060-z

[33] Dulebenets MA. The green vessel scheduling problem with transit time requirements in a liner shipping route with emission control areas. Alexandria Engineering Journal. 2018;57(1):331-342. DOI: 10.1016/j.aej.2016.11.008

[34] Dulebenets MA. A comprehensive multi-objective optimization model for the vessel scheduling problem in liner shipping. International Journal of Production Economics. 2018;196:293-318. DOI: 10.1016/j.ijpe.2017.10.027

[35] Kim DW, Kim KH, Jang W, Chen FF. Unrelated parallel machine scheduling with setup times using simulated annealing. Robotics and Computer-Integrated Manufacturing. 2002;18(3):223-231. DOI: 10.1016/S0736-5845(02)00013-3

[36] Caragiannis I. Efficient coordination mechanisms for unrelated machine scheduling. Algorithmica. 2013;66(3):512-540. DOI: 10.1007/s00453-012-9650-6

A Brief Survey on Intelligent Swarm-Based Algorithms for Solving Optimization Problems

Siew Mooi Lim and Kuan Yew Leong

Additional information is available at the end of the chapter

http://dx.doi.org/10.5772/intechopen.76979

Abstract

This chapter presents an overview of optimization techniques followed by a brief survey on several swarm-based natural inspired algorithms which were introduced in the last decade. These techniques were inspired by the natural processes of plants, foraging behaviors of insects and social behaviors of animals. These swam intelligent methods have been tested on various standard benchmark problems and are capable in solving a wide range of optimization issues including *stochastic*, *robust* and *dynamic* problems.

Keywords: optimization, artificial intelligence, swarm intelligence, nature-inspired and bio-inspired computation

1. Introduction

Optimization is a form of mathematical procedure for determining optimal allocation of scare resources. In recent years, the optimization area has received enormous attention primarily due to the rapid emerging science and technology in computing, communication, engineering, environment and society. Several types of optimization problems exist. Two important classes of objects for most optimization problems are limited resources and activities. Resources include land size, plant capacity and sales force. Whereas production activities are like *produce stainless steel, low carbon steel* or *high carbon steel*; how we solve them will depend on the circumstances to determine the best condition of activity levels using the resources available. All optimization problems have an objective function, constraints, and choice variables which will lead to the improvement in application or audience. For instance, tradeoffs between faster algorithm with more consumption of memory and vice-versa are used to bring the greatest interest to the audience [1].

Three categories of optimizations techniques namely *stochastic optimization* (SO), *robust optimization* (RO) and *dynamic optimization* (DO) are presented in the following subsections with the conclusions given on the advantages and application in practice for each technique. The main motivation behind this study on the nature inspired computation is to identify among the connection, social conduct and rise. This work is needed in the current scientific community to utilize the use of computing to demonstrate the living marvels, to investigate and to enhance our life by using computers. This study will substantially contribute in bringing the inspiration of computerized solutions through a wide range of nature processes.

1.1. Stochastic optimization

Stochastic optimization (SO) process involves randomness in the minimization or maximization of a function and lends itself to real-life phenomena which involve uncertainty and imprecision. The randomness may be present as either noise in measurements or Monte Carlo randomness in the search procedure, or both. Some common techniques of SO are: *direct search methods, stochastic approximation, stochastic programming, simulated annealing, genetic algorithms,* etc. These techniques can cope with the inherent system noise, and systems with high nonlinearity and high dimensional models.

In other words, these models are derived, solved analytically or numerically and analyzed to extract information to be presented to decision makers [2]. SO is important in analyzing, designing and operating modern systems. Specific applications of SO in business include short and long-term investment decisions, aerospace engineering in designing missile or aircraft, new drug design and the network in traffic control. The challenge in real-life applications is hard to estimate the accurate probabilistic description of the randomness, if such information is available, stochastic programming can be applied as a powerful modeling tool. SO has the advantage of solving problems in polynomial time. Theoretically, it guarantees the quality of the solutions generated. Practically, SO is limited by its heavy dependency on the availability of historical data and complex modeling [3, 4].

1.2. Robust optimization

Robust optimization (RO) is a rather new approach that deals with data uncertainty. The two motivational factors of RO are firstly the uncertainty model is rather deterministic and set-based. This motivational concept is the most appropriate notion of parameter uncertainty in many applications. The second motivational factor is the computational tractability. For instance, for a given optimization problem, multiple robust versions exist depend on the structure of the uncertainty set, therefore maintaining tractability is important. The classification models for RO includes local vs. global and probabilistic vs. non-probabilistic. Based on the nature of the problem, this technique is also known as *min-max* or *worst-case* approach. It provides a good guaranteed solution for most possible realizations of the uncertainty in the data. It is also useful if some of the parameters belong to the estimation process and contains estimation errors.

One important concept in defining and interpreting robustness and the resulting models is constraint robustness (model robustness) [5]. The application of RO in engineering is known as *robust design optimization* or *reliability-based design optimization* where the solutions remain feasible for all possible values of the uncertain inputs. RO methodology is applicable to every generic optimization problem in which numerical data can be separated from the structure of the problems. The challenge of RO is that it gives the same weight and values to all of the uncertain parameters. The advantages of RO formulation are cost saving and increment of stability, qualitative and quantitative robustness. The practical usage of RO is that it does not significantly increase the complexity of the considered optimization problems in most cases [6, 7].

1.3. Dynamic optimization

Dynamic optimization (DO), also known as dynamic programming is a process of finding the optimal control profile of one or more control parameters of a system. It is used to find the possible number of solutions for a given problem. There are several approaches of DO such as based on the calculus variations, deal with optimization discrete time and extend the static optimization. Basically, the process of DO implementation involves a system controller, a performance criterion and an algorithm to execute the control. Two key attributes of DO are optimal substructure and overlapping sub-problems [8]. Four major steps on development of DO algorithm are:

a. Characterize the structure of an optimal solution.

b. Recursively define the value of an optimal solution.

c. Compute the value of an optimal solution in a bottom-up fashion.

d. Construct an optimal solution from computed information.

The advantage of this paradigm: it performs the optimization recursively by dividing the problems into a collection of simpler sub-problems. Each sub-problem is solved only once using either top-down or bottom-up approach. To facilitate its lookup, a technique called *memorization* is applied where the solutions of subproblems are indexed based on its input parameter values, thereby solving computation time at the expense of modest expenditure in storage space. Practically, the concept of DO is universal and flexible which can be applied to the execution of any effort [9].

2. Algorithms

Artificial intelligence (AI) has been viewed as a regulation in computer science. It has been developing and examining frameworks which work logically. Bio-inspired computation, metaheuristics and computational intelligence are the common examples of algorithms from numerous parts of AI. Bio-inspired computation utilizes the computing power to demonstrate the living marvels. Computational intelligence which emphasizes on strategy and outcome can be broadly divided into five dominant fields: swarm intelligence, evolutionary computation, artificial neural networks, artificial immune system and fuzzy systems. This chapter will be focusing on a few swarm intelligence-based algorithms which are inspired by their natural processes.

3. Swarm intelligence

Swarm intelligence (SI) is evaluated as an adaptive strategy which takes collective intelligence as a behavior without centralized control structure on how an individual should behave. The rules of SI are simple, self-organizing, co-evolution and being widely applied in the domains of optimizing, searching methods, research in DNA computing improvement, heating system planning etc. SI paradigm includes bird flocking, cuckoo search, animal herding and fish schooling etc. However, the two dominant subfields of SI are ant colony optimization, inspired by pheromone-trail of the ant behavior and particle swarm optimization, inspired by flocking and swarming behavior [10].

However, providing a complete review to all the swarm-based algorithms is rather impossible. The next sub-sections present the inspiration, working, metaphor and heuristic of eight popularly known swarm-based methods. These methods have been introduced and implemented in the last decade. The main challenges of the field and their future trends have also been discussed.

3.1. Bat algorithm

Bat algorithm (BA) [11] helps in simplicity and flexibility. It is found to be very efficient in handling nonlinear and multi objective issues. Bats have a special high-level capability of bio-sonar (echolocation) which is used to find their prey, obstacles, roosting crevices detection and discriminate different types of insects.

The efficiency of BA depends on the features below:

a. Automatic zooming: this capability is performed based on the automatic switch from explorative direction to the local insensitive exploitation.

b. Frequency tuning: the variation of frequency is performed on the echolocation.

Microbats are the famous examples among all the bat species. The echolocation attribute of microbat is used to model BA. Literature has reported a diverse range of BA applications such as loading pattern of nuclear core in engineering optimization, nonlinear economic dispatch problem, design of a power system stabilizer, size optimization for the skeletal structures which consist of truss and frame, multilevel image thresholding which is an image processing technique. In the context of inverse problem and parameter estimation, bat calculations have been utilized in solving numerical improvement, advancing the brushless DC wheel engines, and enhancing topological shape in microelectronic applications [12–15].

There are some successful implementations of BA in SO. In their work of stochastic resonance for MR images enhancement [16], proposed a neuron model that tapped on the BA multi-objective optimization property to tune the parameters. In their work, the BA is utilized to maximize both the image performance indices contrast enhancement factor and the mean opinion score. Their results show that the method has improved the gray-white matter differentiation, which has been found useful to diagnose MR images. In another work by [17], BA is adapted with inclusion of two operations—(1) iterative local search, and (2) stochastic inertial weight to improve its performance in terms of accuracy, speed and convergence stability. It is claimed that BA is easy to fall into local optima and has unstable optimization results due to low global exploration ability. The authors overcome the weaknesses of BA when their iterative local search algorithm disturbs the local optimum and do some local re-search, such that the BA has better ability to get out of the local optima. Adding with their stochastic inertial weight to disrupt the velocity updating equation, it enhances the diversity and flexibility of bat population. They proved their results based on 10 classic benchmark functions, CEC 2005 benchmark suite, and two (2) real-world problems, in which they concluded with improved performance.

A robust tuning of power system stabilizer is demonstrated to be possible by using BA [18]. In such scenario of RO application, the stability of the power system is highly critical. This paper proposed BA to optimize the gain and the pole-zero parameters of the stabilizer. They compared that the BA approach is superior than PSO optimization method. The optimization was performed with objective function based on eigenvalue shifting to guarantee the stability of nonlinear plant for a wide range of operating conditions.

A dynamic perceptive BA [19] is used to optimize particle filter for multiple targets tracking. This is an example of DO in which the authors proposed a multiple-maneuvering-target tracking algorithm and combined it with the BA to optimize particle filter typically used in a modern radar tracking system. Their combined algorithm regards the particles as bats and it simulates the behavior of bats preying by dynamically adjusting the radar tracking system's components of frequency, volume and pulse rate. This dynamic control of adjusting the particle filter, adding with a joint probabilistic data association has enabled an improved accuracy in target tracking even under a complex environment.

In other relevant applications, BA has been reported in data mining techniques of classifications and clustering. BA has been applied in grouping microarray information, minimization of make span and mean flow time to study half breed flow shop booking issues [20]. In the application of image processing, BA has been utilized for full body human stance estimation. In this study, BA has outperformed particle swarm optimization, particle filter and annealed particle filter. Bat based model has also shown its effectiveness in envision coordinating as compared in evolution and genetic algorithms [15]. In fuzzy logic and other applications, BA has been applied in investigating the ideal capacitor position for misfortune decrease in dispersion frameworks. BA and fluffy frameworks have also been utilized for energy displaying and energy changes in a gas turbine [15].

3.2. Firefly algorithm

The current population of firefly species is over 2000. The short and rhythmic blazing light of fireflies is an astounding sight in the sky of the tropical and calm areas. This nature capability of fireflies inspired the firefly algorithm (FA) [21]. Bioluminescence is the process of generating the flash light. The light is used to model the warning signals. Each objective function of optimization problem is represented by different light intensity. They are some similarities between FA and bacterial foraging algorithm. Their attractions are based on objective function, fitness and distance respectively. FA can solve discrete and continuous optimization problems. Kwiecien and Filipowicz [22] have applied FA in cost optimization of queueing systems. A study carried out by Gandomi et al. [23] has proven that FA is better than other nonlinear optimization techniques in designing the stepped cantilever beam. FA has been reported in the recent literature as an efficient computational procedure for simultaneously generating multiple different alternatives to an optimal solution [24].

FA has a wide spread of applications since its introduction, attributed mainly to its simplicity of implementation as compared to some traditional approaches. In content-based image retrieval, feature extraction has been done with the Euclidean distance estimation between the pixels. However, such approach needs more precision, and this has motivated [25] to employ FA to optimize the image features. Their work is closely related to SO as the potential image features are stochastically found by FA. They benchmarked their FA image optimization results with PSO and GA and discovered the differences of each model in terms of precision and image recall. [26] also applied FA to solve a SO problem in linear phase finite impulse response filter (FIR) design. Differential evolution (DE) is known as one of the best performer in used for such problem. However, they proved through their simulation of designing FIR filters that FA is better than other relevant algorithms (inclusive of PSO and GA). The improvement was recorded not only in the convergence speed but also in the performance of the designed filter.

Other variants of FA have also found their applications in several various disciplines. [27] proposed a hybrid PSO-FA to solve a combinatorial optimization issue in floor planning. [28] introduced a hybrid FA and DE method to estimate the parameters of the nonlinear biological model. In the optimization design of sewer pipes, [29] has proposed a novel method by combining a support vector regression and the FA to predict the minimum velocity required to avoid sediment settling in the pipe channels.

3.3. Lion optimization algorithm

Lion optimization (LO) [30] is a population-based algorithm which was inspired by lion's social system and collaboration characteristics which can be described with the term 'pride'. The uniqueness of lion's social behavior makes them the strongest mammal in the world. LO is modeled based on two unique behaviors of lion: territorial defense and territorial takeover. Based on these two behaviors, the solutions of LO are generated through three steps: (1) to differentiate whether each cub solution is an original or a derived solution; (2) territorial defense will proceed to evaluate and compare the existing and new solutions; (3) if the existing solution is better than the new solution, then territorial takeover will keep the existing solution to further improve it. LO can perform huge search space to solve continuous, single variable and multi-variable optimization problems. LO algorithm has been validated using De-Jong's Type 1 function and the performance has been compared against evolutionary programming. The results shown that LO performed better than evolutionary programming.

LO is still in its early stage of application, both [31, 32] have experimented on its optimization capability and benchmarked it on some function optimization problems. Both authors concluded on the high performance of LO in functions optimization. While [33] has taken LO further to perform clustering of data by utilizing the optimization capability of LO. Such data clustering approach is very much a SO problem. In their work, the LO is modified with the fractional theory to search the cluster centroids of data instead of typical distance measurement.

3.4. Chicken swarm optimization algorithm

Chickens are sociable birds that they live in groups. Chicken swarm optimization (CSO) algorithm [34] was inspired by the social behavior of chickens. Every chicken has their own motion laws. The fitness values of the chickens will identify themselves as the roosters, hens and chicks. These identities will put them into groups. The fittest, the weakest and the intermediate chickens are simulated as a rooster, as a chick and as hens accordingly. The hierarchal order plays a vital part in the group of chickens and this characteristic is used to model CSO algorithm. CSO has been applied in solving the design of speed reducer efficiently. In that research, a gearbox has been created with the design of most efficient speed. The research on CSO has been promising. It has been used to improve the performance of the greedy algorithm [35].

A deadlock-free migration of virtual machine consolidation is optimized using CSO [36]. In such consolidation of services, two separate but related issues of virtual machine placement and migration is a challenging optimization task. The authors proposed a consolidation scheme utilizing CSO that turns the virtual machine consolidation problem into a vector packing optimization based on deadlock-free migration. The optimization also helps to minimize the energy consumptions. The proposed method achieved higher convergence rate as compared to several other deadlock-free migration algorithms.

CSO is also applied to a classical job-shop scheduling problem in the work of [37]. An improved version of CSO is also being applied to identifying the maximum power point tracking control of a photovoltaic system [38]. CSO has also seek its application in disaggregation of non-invasive domestic appliances [39].

3.5. Social spider algorithm

Social spiders are organism living in groups. They are solitary and having aggressive characters among their own species. Their foraging behaviors and their corporation in performing daily tasks are used to model social spider algorithm (SSA) [40]. In SSA, two different evolutionary operators are created based on the gender of male and female spiders to divide their tasks for predation, web design and mating. This algorithm can solve a wide range of continuous optimization problems including minimization of molecular potential energy function [41, 42]. SSA has been validated using standard benchmark problems to study its performance. An analysis has been carried out on the performance SSA against particle swarm optimization and artificial bee colony. The results shown that SSA has outperformed the other techniques.

SSA has some wide application in recent researches. An SSA is proposed to solve a non-convex economic load dispatch problem [43]. Economic load dispatch (ELD) is one of the essential components in power system control and operation. Most modern power system introduces new models of the power units which are non-convex, non-differentiable, and sometimes non-continuous; such RO problem is hence difficult to be solved by conventional mathematical techniques. In this paper, the authors modified the SSA to suit the characteristics of ELD, and their simulation results show that such ELD problem can be solved by SSA effectively and efficiently.

Another example of employing SSA to a DO problem is proposed by [42] to solve the transmission expansion planning in electrical power system. The authors tested SSA in solving the transmission expansion planning problem of three benchmark systems having 6-busses, 46-busses, and 87-busses. They achieved great performance as well as reduction in the total investment cost.

In a multi-objective optimization problem of QoS-aware web services, [44] applied SSA to perform optimized selection of numerous functionality in the web services involving complex tasks delivery. They studied current approaches of GA and PSO to realize that the time performance of such approaches is still a great concern. The proposed SSA has outperformed PSO in terms of both execution time and fitness.

3.6. Spider monkey optimization algorithm

Spider monkey (SM) [45] is a population-based optimization algorithm which was inspired by the intelligent ways of spider monkeys to search for the most suitable food sources. The excellency of food source corresponds to the fitness of a solution. Major characteristics and the strategies of SM algorithm are similar to artificial bee colony algorithm. The exploration and exploitation functions allowed SM algorithm to perform huge search space and generate greatest feasible solutions. SM algorithm is simple and speedy. It is used to solve numerical optimization problems.

An example of SM application in continuous numerical optimization can be found in the work of [46]. They modified SM to enable it for solving some constrained optimization problems. Their proposed SM was tested on the well-defined constrained optimization problems of CEC2006 and CEC2010 benchmark sets. The algorithm acquired some promising results when compared to PSO, artificial bee colony and DE methods.

Although SM could be best to implement for a numerical problem, several other researchers have applied it to some wider scope of optimization problems. Both [47, 48] in their separate studies, have proposed SM to solve antenna optimization. [47] focuses their work in the thinning of concentric circular antenna arrays. While such thinning problem is a binary optimization, the original SM might not be suitable. Hence, they suggested their version of binary SM in which it must handle the logical operators of the thinning problem. Their results proved the competence and superiority of a binary SM as compared to existing metaheuristic algorithms. On the other hand, [48] applies SM to synthesize the factor of a linear antenna's array and to optimally design an E-shaped patch antenna. They discovered that their SM, when compared to traditional method for such optimization problem, can reach optimum solutions with less number of iterations.

3.7. African buffalo optimization algorithm

African buffalo optimization (ABO) algorithm [49] was inspired by the practice of African buffalos in the vast African forests and savannahs in finding pastures. Three specific characteristics of these animals are used to model ABO algorithm. Firstly, these animals have high memory capacity. This skill enables them to monitor their routes up to thousands of kilometers in Africa continent. Secondly, they communicate among themselves using two specific vocalizations: 'maa' and 'waa' to support each other in surviving. Lastly, African buffalos practice 'democracy' system in making decisions.

The information of every buffalo's past and current location is used in tackling the issue of premature convergence. Leading buffalo's search space and the experience of all other buffalos complement to the exploration and exploitation strategies in ABO. This algorithm utilizes only learning parameters, therefore it is a simple and yet easy to implement algorithm which guarantees quick convergence. The efficiency and powerful features of this algorithm are capable in solving knapsack problems. ABO has been validated using traveling salesman benchmark problems to study its cost effectiveness. A study has been carried out on the comparative CPU time for ABO against Genetic Algorithm, Honey Bee Mating Optimization, Ant Colony Optimization, Simulated Annealing and Adaptive Simulated Annealing with Greedy Search. ABO has outperformed all other techniques to obtain the solutions at an incredibly fast rate.

[50] has used ABO to solve the well-studied Traveling Salesman's Problem (TSP). They performed ABO on 33 benchmark symmetric TSP and observed excellent exploration and exploitation of the search space through regular communication, cooperation, good memory of its previous individual and collective exploits. They concluded that ABO is as competitive as other algorithms superior in TSP.

In another DO example, ABO is used to optimize parameters tuning of proportional-integral-derivative (PID) controller [51]. The PID controller in their study is used to automatically regulate voltage. It was noted that existing metaheuristic tuning methods have been proven to be quite successful, but the authors wanted to improve the gain overshoot and steady state errors of the system. They gained some encouraging results with ABO when it was compared to several other optimization algorithms.

3.8. Flower pollination algorithm

The goal of plants, like any other living organism is producing offspring for the next generation. Pollination is the process of transferring pollen grains from the male anther of a flower to the female stigma. Two types of pollinations process are self-pollination and cross pollination. When pollen grains are produced, pollinators will spread it among the flowers to either local or global flow of pollination. The process of passing the pollen grains from the stamens to the ovule-bearing organs during pollination is used to model flower pollination algorithm (FPA) [52].

The assumption on the current version of FPA: each flower only produces one pollen gamete. The time complexity of FPA is shorter; therefore, it is flexible and easy to implement. A study has been carried in structural engineering to evaluate the cost optimization of tubular column under compressive load using FPA against Cuckoo search algorithm, fuzzy rules and engineering optimization techniques. The results showed that FPA is the most efficient method and the convergence of the algorithm is very effective. This algorithm has been applied for solving continuous, single objective and multi-objective optimization problems. It can be improved further to apply in the areas of image compression and graph coloring. FPA has successfully converged in shape matching problem based on a relatively new branch called atomic potential matching model [53].

FPA is applied to a visual tracking problem in the work of [54]. In their model, visual tracking is considered to be a process of optimal reproduction of flowering plants. This is a typical example of DO. FPA is then presented with a switch probability that changes dynamically with generation numbers. To compare the tracking ability of the FPA tracker, the tracking accuracy of particle filter, mean shift and PSO are studied as well. Comparative results show that their method outperforms the other three trackers. Other applications of FPA can be found in the works of [55–57]. [55] proposed a FPA to optimize a SO problem of static economic dispatch incorporating wind farm. [56] worked on the sizing optimization of truss structures. Again, this is another example of SO. While [57] applied FPA on a DO problem of photovoltaic parameters optimization. Through their simulation, FPA is recommended as the fastest and the most accurate optimization technique for the optimal parameters extraction process, after benchmarking it on several other methods.

4. Conclusion

All different creatures survive with their own unique behaviors and features. Their characteristics are concealing in the natural world. This chapter has reviewed eight natural inspired algorithms mainly from the field of swarm intelligence. These techniques are becoming powerful in numerical optimization and have shown remarkable robustness, high accuracy and have immense capacity in solving different types of optimization problems.

In dealing with real world problems of optimization, we are now presented with even more choices of algorithm thanks to inspirations form the nature, as well as persistence research contributions by many researchers. However, with more choices means it could be even harder to decide which one of the algorithms is a better candidate for a problem at hand. Based on No-Free-Lunch theorem, there is surely no single best algorithm for every problem. The complexity, characteristics and diversity of optimization problems mean that it is very unlikely to have a single method that can handle all types of optimization problems. This is very much the current state of research in optimization, despite the abundance of nature inspired algorithms being added to the solution pool.

Moving ahead, the research community is definitely not looking for a single most powerful algorithm to solve all types of problem. Indeed, we might witness even more successful variants of algorithm being developed from their current counterparts. This is a trend we have observed since any introduction of a new algorithm being inspired. It is also possible that more and more new optimization algorithms are to be inspired, given that the vast unknown and untested phenomena in the nature are still beyond our exploration.

Author details

Siew Mooi Lim[1]* and Kuan Yew Leong[2]

*Address all correspondence to: limsm66@gmail.com

1 Faculty of Computing, University Malaysia of Computer Science and Engineering, Putrajaya, Malaysia

2 School of Information Technology (Caulfield), Monash University, Melbourne, Victoria, Australia

References

[1] Crowley M. Using equilibrium policy gradients for spatiotemporal planning in forest ecosystem management. IEEE Transactions on Computers. 2014;**63**(1):142-154

[2] Weisstein EW. Stochastic optimization. From MathWorld—A Wolfram Web Resource

[3] Mesbah A. Stochastic model predictive control: An overview and perspectives for future research. IEEE Control Systems. 2016;**36**(6):30-44

[4] Schmidhuber J. Deep learning in neural networks: An overview. Neural Networks. 2015; **61**:85-117

[5] Dunning I, Huchette J, Lubin M. JuMP: A modelling language for mathematical optimization. SIAM Review. 2017;**59**(2):295-320

[6] Ahmadi-Javid A, Jalali Z, JKlassen K. Outpatient appointment systems in healthcare: A review of optimization studies. European Journal of Operational Research. 2017;**258**(1):3-34

[7] Bertsimas D, Gupta V, Kallus N. Data-driven robust optimization. Springer Link. 2018; **167**(2):235-292

[8] Shin YC, Xu CY. Intelligent Systems: Modelling, Optimization and Control. CRC Press, Taylor & Francis Group; 2017

[9] Priya Esther B, Sathish Kumar K. A survey on residential demand side management architecture, approaches, optimization models and methods. Renewable and Sustainable Energy Reviews. 2016;**59**:342-351

[10] Brownlee J. Clever Algorithms: Nature-Inspired Programming Recipes. 2012. ISBN: 978-1-4467-8506-5

[11] Yang X-H. A new metaheuristic bat-inspired algorithm. In: Gonzalez JR et al. Nature Inspired Cooperative Strategies for Optimization (NICSO 2010), SCI. Vol. 284. 2010. pp. 65-74

[12] Tsai PW et al. Bat algorithm inspired algorithm for solving numerical optimization problems. Applied Mechanics and Materials. 2011;**148-149**:134-137

[13] Titus J. Careful Designers Get the Most from Brushless DC Motors. Tillgänglig: ECN Electronic Communication Network; 2012

[14] Yang X-S. Bat algorithm: Literature review and applications. International Journal of Bio-Inspired Computing. 2013;**5**(3):141-149

[15] Marichelvam MK, Prabaharam T. A bat algorithm for realistic hybrid flow shop scheduling problems to minimize makespan and mean flow time. ICTACT, Journal on Soft Computing. 2012;**3**(1):428-433

[16] Singh M, Verma A, Neeraj S. Bat optimization-based neuron model of stochastic resonance for the enhancement of MR images. Biocybernetics and Biomedical Engineering. 2017;**37**(1):24-134

[17] Gan C, Cao WH, Wu M, Chen X. A new bat algorithm based on iterative local search and stochastic inertia weight. Expert Systems with Applications. 2018;**104**:202-212

[18] Sambariya DK, Prasad R. Robust tuning of power system stabilizer for small signal stability enhancement using metaheuristic bat algorithm. International Journal of Electrical Power and Energy Systems. 2014;**61**:229-238

[19] Chen ZM, Bo YM, Tian MC, Wu PL, Ling XD. Dynamic perceptive bat algorithm used to optimize particle filter for tracking multiple targets. Journal of Aerospace Engineering. 2018;**31**(3)

[20] Yang X-H. A new metaheuristic bat-inspired algorithm. In: Gonzalez JR et al., editors. Nature Inspired Cooperative Strategies for Optimization (NICSO 2010), SCI. Vol. 284. 2010. pp. 65-74

[21] Bacanin N, Tuba M. Firefly algorithm for cardinality constrained mean-variance portfolio optimization problem with entropy diversity constraint. The Scientific World Journal. 2014. Article ID 721521, http://dx.doi.org/10.1155/2014/721521

[22] Kwiecien J, Filipowicz B. Firefly algorithm in optimization of queueing systems. Bulletin of the Polish Academy of Sciences: Technical Sciences. 2012;**60**(2):363-368

[23] Gandomi AH, Yang X-S, Alavi AH. Mixed variable structural optimization using firefly algorithm. Computers & Structures. 2011;**89**(23–24):2325-2336

[24] Yeomans JS. An efficient computational procedure for simultaneously generating alternatives to an optimal solution using the firefly algorithm. In: Yang XS, editor. Nature-Inspired

Algorithms and Applied Optimization. Studies in Computational Intelligence, Vol. 744. Springer; 2018

[25] Kanimozhi T, Latha K. Stochastic firefly for image optimization, International Conference on Communication and Signal Processing. 2013. pp. 592-596

[26] Suman S, Kar R, Mandal D, Ghoshal S. A novel firefly algorithm for optimal linear phase FIR filter design. International Journal of Swarm Intelligence Research (IJSIR). 2013;**4**(2):29-48

[27] Sivaranjani P, Senthil Kumar A. Hybrid particle swarm optimization-firefly algorithm (HPSOFF) for combinatorial optimization of non-slicing VLSI floor planning. Journal of Intelligent and Fuzzy Systems. 2017;**32**(1):661-669

[28] Afnizanfaizal A, Safaai D, Sohail A, Arjunan SNV. An evolutionary firefly algorithm for the estimation of nonlinear biological model parameters. PLoS ONE. 2013;**8**(3):56310

[29] Ebtehaj I, Bonakdari H. A support vector regression-firefly algorithm-based model for limiting velocity prediction in sewer pipes. Water Science and Technology: A Journal of the International Association on Water Pollution Research. 2016;**73**(9):2244-2250

[30] Rajakumar BR. The lion's algorithm: A new nature-inspired search algorithm. Procedia Technology. 2012;**6**:126-135

[31] Yazdani M, Fariborz J. Lion optimization algorithm (LOA): A nature-inspired metaheuristic algorithm. Journal of Computational Design and Engineering. 2016;**3**(1):24-36

[32] Wang B, Jin XP, Cheng B. Lion pride optimizer: An optimization algorithm inspired by lion pride behaviour. Science China Information Sciences. 2012;**55**(10):2369-2389

[33] Chander S, Vijaya P, Dhyani P. Fractional lion algorithm-an optimization algorithm for data clustering. Journal of Computer Science. 2016;**2**(7):323-340

[34] Meng X, Liu Y, Gao XZ, Zhang H. A new bio-inspired algorithm: Chicken swarm optimization. In: Advances in Swarm Intelligence. ICSI, Lecture Notes in Computer Science. Vol. 8794. Springer; 2014. pp. 86-94

[35] Mohamed TM. Enhancing the performance of the greedy algorithm using chicken swarm optimization: An application to exam scheduling problem. Egyptian Computer Science Journal. 2018;**42**(1):1-17

[36] Tiana F, Zhanga R, Lewandowskic J, Chaoc KM, Li LZ, Dong B. Deadlock-free migration for virtual machine consolidation using chicken swarm optimization algorithm. Journal of Intelligent & Fuzzy Systems. 2017;**32**(2):1389-1400

[37] Xu S, Wu D, Kong F, Ji Z. Solving flexible job-shop scheduling problem by improved chicken swarm optimization algorithm. Journal of System Simulation. 2017;**29**(7):1497-1505

[38] Benoit C, Sebastian S, Wu ZQ, Yu DQ, Kang XH. Application of improved chicken swarm optimization for MPPT in photovoltaic system. Optimal Control Applications and Methods. 2018;**39**(2), 1029(14)

[39] Xu Y, Li W, Li D, You X. Disaggregation for non-invasive domestic appliances based on the improved chicken swarm optimization algorithm. Power System Protection and Control. 2016;**44**(13):27-32

[40] Yu JJQ, Li VOK. A social spider algorithm for global optimization. Applied Soft Computing. 2015;**30**:614-627

[41] Tawhid MA, Ali AF. A hybrid social spider optimization and genetic algorithm for minimizing molecular potential energy function. Soft Computing. Springer; 2017;**21**(21):6499-6514

[42] El-bages MS, Elsayed WT. Social spider algorithm for solving the transmission expansion planning problem. Electric Power Systems Research. 2017;**143**:235-243

[43] Yu JJQ, Li VOK. A social spider algorithm for solving the non-convex economic load dispatch problem. Neurocomputing. 2016;**171**:955-965

[44] Mousa A, Bentahar J. An efficient QoS-aware web services selection using social spider algorithm. Procedia Computer Science. 2016;**94**:176-182

[45] Bansal JC, Sharma H, Jadon SS, Clerc M. Spider monkey optimization algorithm for numerical optimization. Memetic Computing. 2014;**6**(1):31-47

[46] Gupta K, Deep K, Bansal J. Spider monkey optimization algorithm for constrained optimization problems. Soft Computing. 2017;**21**(23):6933-6962

[47] Singh U, Salgotra R, Rattan M. A novel binary spider monkey optimization algorithm for thinning of concentric circular antenna arrays. IETE Journal of Research. 2016;**62**(6):736-744

[48] Al-Azza A, Al-Jodah A, Ammar F, Harackiewicz J. Spider monkey optimization: A novel technique for antenna optimization. IEEE Antennas and Wireless Propagation Letters. 2016;**15**:1016-1019

[49] Odili JB, Kahar MNM, Anwar S. African buffalo optimization: A swarm-intelligence technique. Procedia Computer Science. 2015;**76**:443-448

[50] Odili J, Beneoluchi M, Kahar MN. Solving the traveling Salesman's problem using the African buffalo optimization. Computational Intelligence and Neuroscience. 2016;**2016**(3)

[51] Odili J, Noraziah. Parameters-tuning of PID controller for automatic voltage regulators using the African buffalo optimization. PLoS One. 2017;**12**(4):175901

[52] Yang X-S. Flower pollination algorithm for global optimization. In: Unconventional Computation and Natural Computation, Lecture Notes in Computer Science. Vol. 7445. 2012. pp. 240-249

[53] Zhou Y, Zhang S, Luo Q, et al. Neural Computing and Applications. 2018;**29**:21. DOI: 10.1007/s00521-016-2524-0

[54] Gao ML, Jin S, Jun J. Visual tracking using improved flower pollination algorithm. Optik - International Journal for Light and Electron Optics. 2018;**156**:522-529

[55] Velamuri, Suresh S, Sreejith P, Ponnambalam. Static economic dispatch incorporating wind farm using flower pollination algorithm. Perspectives in Science. 2016;**8**:260-262

[56] Gebrail B, Melih NS, Yang XS. Sizing optimization of truss structures using flower pollination algorithm. Applied Soft Computing. 2015;**37**:322-331

[57] Alam DF, Yousri DA, Eteiba MB. Flower pollination algorithm based solar PV parameter estimation. Energy Conversion and Management. 2015;**101**:410-422